ALTERNATIVE ECONOMIC STRATEGIES

For Maria and Stephanos Tsakalotos

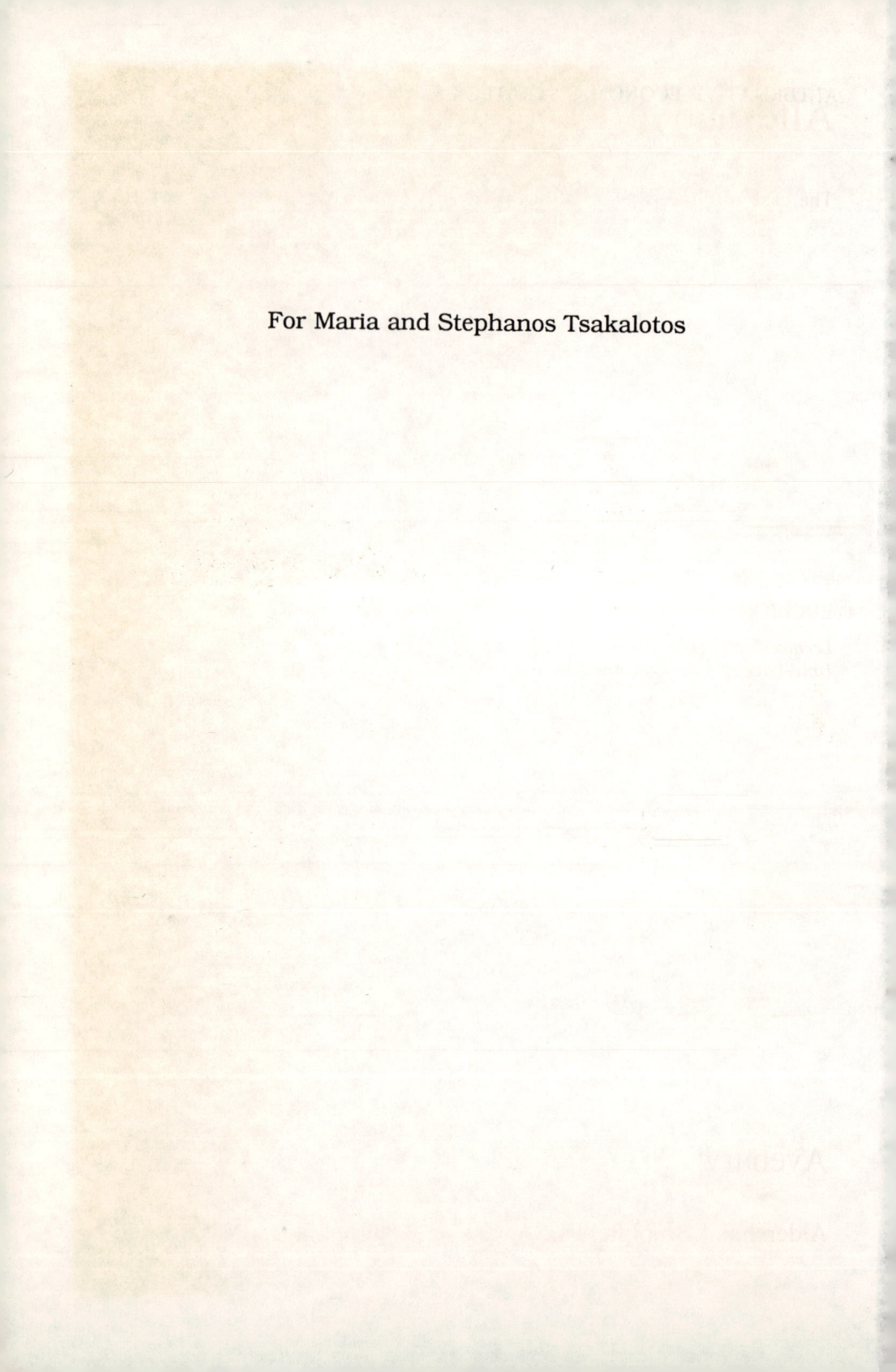

Alternative Economic Strategies

The case of Greece

EUCLID TSAKALOTOS
Lecturer in Economics
University of Kent at Canterbury

Avebury

Aldershot · Brookfield USA · Hong Kong · Singapore · Sydney

© Euclid Tsakalotos 1991

All rights reserved. No part of this publication may be reproduced, stored in a retrieval system, or transmitted in any form or by any means, electronic, mechanical, photocopying, recording or otherwise without the prior permission of the publisher.

Published by
Avebury
Academic Publishing Group
Gower House
Croft Road
Aldershot
Hants GU11 3HR
England

Gower Publishing Company
Old Post Road
Brookfield
Vermont 05036
USA

HC
295
.T683
1991

A CIP catalogue record for this book is available from the British Library and the US Library of Congress.

ISBN 1 85628 183 3

Printed in Great Britain by Athenaeum Press Ltd, Newcastle upon Tyne.

Contents

ACKNOWLEDGEMENTS ix
ABBREVIATIONS xi
INTRODUCTION xiii

Chapter 1: Alternative Economic Strategies: Socialism, Institutional Change and Economic Policy 1

    1.1 The revolutionary path to socialism 3
    1.2 Social democracy and socialism 6
    1.3 The economics of social democracy 11
        1.3.1 Economics and the social democratic consensus 12
        1.3.2 Left social democracy and political economy 19
    Notes 30

Chapter 2: Indicative Planning 36

    2.1 The economic theory of indicative planning 37
    2.2 The French model 43
    2.3 The political economy of indicative planning 50
    Notes 58

Chapter 3: Planning and Macroeconomic Control — 60

- 3.1 The national plan and the scope of planning — 61
- 3.2 Beyond indicative planning — 66
  - 3.2.1 Public and social ownership — 67
  - 3.2.2 Enterprise boards and new public enterprises — 72
  - 3.2.3 Planning agreements, financial incentives and commercial leverage — 82
- 3.3 A complementary framework for planning — 86
  - 3.3.1 Financial institutions — 87
  - 3.3.2 Trade and capital flows — 88
  - 3.3.3 Competition policy — 90
- 3.4 Industrial democracy and planning — 91
- 3.5 The coherence of the planning alternative — 95
- 3.6 The political economy of incomes policy — 102
- 3.7 Social contracts, incomes policies and planning — 107
- 3.8 Conclusions — 114
- Notes — 118

Chapter 4: PASOK's Economic Alternative — 124

- 4.1 PASOK's political strategy — 125
- 4.2 PASOK's economic strategy — 129
  - 4.2.1 Socialisation — 130
  - 4.2.2 Democratic planning — 133
  - 4.2.3 Active planning and industrial strategy — 135
  - 4.2.4 Short-run economic policy and structural change — 138
- 4.3 Conclusions — 141
- Notes — 145

Chapter 5: Macroeconomic Policy and PASOK's Alternative Economic Strategy — 148

- 5.1 Macroeconomic imbalances in 1981 — 149
- 5.2 PASOK's initial economic policy — 152
- 5.3 'Stabilisation through development and gradual adjustment' — 156

	5.3.1	The rationale of 'stabilisation through development'	156
	5.3.2	The underlying economic theory of 'stabilisation through development'	158
	5.3.3	The policy of gradual adjustment	162
5.4	An assessment of the gradual adjustment strategy		166
5.5	Conclusions		178
Notes			183

Chapter 6: Democratic Planning, the Public Sector and Socialisation — 190

6.1	Democratic planning		191
	6.1.1	The 1983-87 plan process	192
	6.1.2	The main goals of the 1983-87 plan	193
	6.1.3	Co-ordination and participation in the plan	195
	6.1.4	Plan consistency	199
6.2	Public investments and 'active' planning		201
	6.2.1	Public investments	201
	6.2.2	Active planning	205
6.3	Planning, socialisation and the DEKO		211
	6.3.1	Economic and organisational problems of the DEKO	212
	6.3.2	The socialisation framework of Law 1365/83	214
	6.3.3	The implementation of the socialisation policy	216
	6.3.4	Socialisation, labour participation and the ASKE	218
	6.3.5	Last attempts at a DEKO policy	223
6.4	Conclusions		224
Notes			227

Chapter 7: Industrial Strategy and the Problematic Firms — 230

7.1	The structural problems of Greek industry and the 'ailing industries' phenomenon	231

7.2	Supervisory councils and sectoral planning		235
	7.2.1	Supervisory councils	235
	7.2.2	Sectoral planning	237
	7.2.3	Assessment of the sectoral industrial strategy	239
7.3	Industrial intervention and the problematic firms		242
	7.3.1	Law 1386/83	242
	7.3.2	The co-ordination of the OAE strategy	245
	7.3.3	The Business Reconstruction Organisation (OAE)	252
7.4	Conclusions		260
Notes			263

Chapter 8: Conclusions: The Political Economy of Alternative Economic Strategies 266

8.1	Planning, state and society		269
	8.1.1	Planning and society	269
	8.1.2	Planning and the state	274
	8.1.3	Institutions and mechanisms of planning	278
8.2	Macroeconomic policy and structural change		282
8.3	Methodological issues		286
Notes			288

Bibliography 291

Acknowledgements

This book originated as a DPhil thesis submitted at the University of Oxford in 1989.

My greatest debt is to my parents whose financial support and encouragement throughout the period of my education has gone far beyond the call of duty.

I am also indebted to Professor W Brus who supervised the whole of the thesis and who had the thankless task of helping me discipline and organise my thoughts.

I am also grateful to the George Webb Medley Fund for providing the financial support for my research in Greece. The research was partly based on a number of interviews with people who gave their time generously. In particular, I would like to thank: Dimitris Halikias, Georgos Kandalepas, Louka Katseli, Michalis Spourdalakis, Kostis Bizanis, Panor Skartsolias, Tasos Tsapatsaris, Dimitris Marinos-Kouris, Stelios Pappas, Georgos Oikonomou, Nikos Manolopoulos, Leutheris Papagiannakis, Loukas Athanasiou, Makis Giakoumelos, Nestor Augoustiniatos, Giannis Drosos, Eli Hadjiconstandi, Isaac Sambethai, Dimitis

Zachariadis, and Aris Karotinos. Zoe Georgiou, Georgos Samiotis, Kostas Zacharis and Amalia Dagli all ensured that my time in Greece was most enjoyable.

I will always remember the guidance and kindness of Keith Perry during the earlier part of my education as well as the friendship of Owen Tudor.

I would like to thank Giannis Stournaras for his detailed reading and comments of this work and the time he gave for our countless discussion. His grasp of both the technical, as well as wider, issues of economics has been a constant source of inspiration.

I will always be grateful for the friendship of Jonathan Davis. His intellectual seriousness and rigour, his love of theoretical debate and his patience have taught me often more than I was willing to learn.

Finally, I will never forget the advice, companionship and support of Heather Gibson. Apart from making detailed comments throughout the writing of this book, she, more than anybody, did more to prevent me 'making a drama out of every crisis'.

Abbreviations

AES	Alternative Economic Strategy
ASKE	Representative Assembly of Social Control
CGP	Comissariat General Du Plan (French Planning Commission)
DEA	Department of Economic Affairs
DEH	Greek Public Power Corporation
DEKO	Public Corporations and Entities
DTI	Deparment of Trade and Industry
EAP	Committee for Analysis and Planning
EPSAS	Committee of Planning Agreements and Development Contracts
ESAP	National Council for Development
ESEP	Executive Council for the Implementation of the Five-Year Plan
GEPI	Italian State Holding Company for intervention in medium-sized and smaller firms
GSEE	Greek Trade Union Federation
IOC	Industrial Organisation Corporation
IRI	Italian State Holding Company
KEPE	Centre for Planning and Economic Research
LEB	Local Enterprise Board

MNC	Multinational Corporation
MNE	Ministry of National Economy
NBPI	National Board for Prices and Incomes
NEB	National Enterprise Board
NEDC	National Economic Development Council
NEDO	National Economic Development Office
NIB	National Investment Bank
NPE	New Public Enterprise
OAE	Business Reconstruction Organisation
OSD	Integrated Networks of Activities
OSE	Greek Public Transport Corporation
OTE	Greek Telecommunications Corporation
PASOK	Panhellenic Socialist Movement
PCF	Communist Party of France
PCI	Communist Party of Italy
PF	Problematic Firm
SC	Supervisory Council
SEB	Confederation of Greek Industrialists
SHC	State Holding Company
SPD	German Social Democratic Party

Introduction

This book looks at alternative economic strategies and takes the Greek case as a test case of such strategies. A first question that arises is alternative to what?

In the post-war period the economics profession was both relatively prestigious and united on the means and ends of economic science. The goals of economic science - full employment, price stability, growth and a stable balance of payments - were widely shared. This was the era of the Keynesian social democratic consensus of full employment, growth and an expanding world economy. Economic arguments revolved around the technical means to arrive and sustain these goals.

By the mid-1960s this consensus was already beginning to look precarious. Although at first monetarism seemed to encompass no major challenge to the previous arrangements, since this was obscured by questions of the significance of the money supply and the relative merits of monetary and fiscal policy in demand management, it soon became clear that a more significant and permanent shift in economic thinking was taking place. It became apparent that monetarism was not merely a different approach to achieving the same goals, but was challenging the means and ends of economic policy. The normative-positive distinction in economics was becoming less secure, as monetarists argued about the superiority of markets not only on economic grounds but as part of a wider argument over democracy, freedom and, more generally, the type of society people should aspire to.

There was a similar reaction from the left of the political spectrum. The shift to the left in the British Labour Party, the Common Programme of the left in France and the rise of the socialist parties in Southern Europe signposted this. Here too the means and ends of economic policy of the social democratic consensus were being challenged. Firstly, it was felt that previous policies had not gone far enough in promoting equality, diminishing poverty and transforming the relations of economic power. There was a greater awareness of the need to treat the economy as a system with specific structural properties which limited the achievement of socialist goals. Furthermore it seemed that the existing set of economic policies - demand management, indicative planning and incomes policies - were no longer even capable of preserving the gains that had been made. Finally, there was a greater interest in, and readiness to employ, more radical economic measures in order to promote socialist values. Questions of planning and state intervention, industrial democracy and extending democracy to the economic sphere returned to the political agenda. This entailed a considerable widening of the scope of economic policy.

It is in this sense that we are entitled to speak of alternative economic strategies. This book makes three contributions. Firstly, it offers a critical analysis of the literature on such alternative economic strategies, on planning and the economics of intervention. Secondly, it examines the applicability of such a model to a country like Greece. Thirdly, by taking the Greek experiment as a test case, it considers the coherence of the alternative model of economic policy by examining the lessons from the Greek case.

The question arises over whether the Greek experience in the 1980s, constitutes a useful test case for 'alternative economic strategies'? PASOK (the Panhellenic Socialist Movement) was formed in 1974 and its conception of itself was that of a radical party committed to a socialist transformation of Greece. Indeed throughout the 1970s it was keen to distinguish itself from what it considered the 'reformist' social democratic parties of Northern

Europe. And as we show in chapter four by October 1981, when it was elected to government, its economic policy, as expressed in its manifesto, party documents and the government programme (presented to parliament in late 1981) was clearly 'alternative' in the sense described above.

There is of course a real danger in taking the pronouncements of any political party at face value. PASOK is hardly the first party of the left whose record in government is somewhat different from its election programme. Furthermore in trying to investigate whether PASOK had a coherent economic strategy by 1981 (one which we call 'alternative' in chapter four) there is the added danger of imposing a rationality, or a unity of purpose, that simply did not exist. It would be naive to think that election manifestos and economic programmes are simply formulated by political parties as guidelines to intended practice. For their role is clearly also to mobilise popular support for the party and for this reason alone economic programmes cannot be simply read as future intentions. While it is important to be fully aware of such dangers, it is difficult to know precisely how to cope with them. And in any case one must avoid the opposite pitfall of prejudging the issue - of taking as given issues which need to be shown.

Nevertheless, the above dangers gain much credence given that we now know the eventual fate of PASOK's alternative economic strategy in the 1980s. But there are dangers too in making too much use of hindsight and implying that a failure was somehow predetermined. To see this we begin by dividing PASOK's spell in government (1981-89) into three periods which is useful for analytical purposes. The first lasted from October 1981 to the election in the summer of 1985 when PASOK was re-elected for a second term. It is in this period that PASOK was most clearly committed to an alternative economic strategy, at least in terms of political rhetoric - the political practice remains to be seen. It is an analysis of this period which constitutes the central concern of the second half of this book. The second period begins with the adoption of the stabilisation programme in October 1985 and ends with the virtual abandonment of this programme in late 1987 in favour of a much more loose ('populist') strategy which had more than one eye on the coming election. While in this second period there was clearly a shift towards a more orthodox economic policy, one should not draw too strong a distinction with the previous period. As we shall see PASOK was still committed to many of the same policies of the previous administration (including the national plan and a dirigiste industrial policy) and this was tied to a strong modernising streak at the Ministry of National Economy (see chapter six). However, in practice, while the 1985-87 stabilisation package was relatively successful, at least in its own terms, little was achieved with those elements of PASOK's policy which could in any way be described as alternative. The final period of 1987-89 is characterised by the demise of PASOK and here it is difficult to detect any rational economic policy whatsoever, whether orthodox or alternative. Furthermore this period was also characterised by serious financial and political scandals which have left a large stigma on the PASOK period as a whole. Overall the PASOK experiment was not a happy one.

Given the above what can be gained from a detailed examination of the PASOK experiment, especially during the first period? There are in fact a number of reasons why such an examination is warranted. Whatever the specificity of the Greek case, there is clearly something to be learnt from any alternative economic strategy, however incomplete its implementation. Furthermore, as we have indicated, one must be wary of explanations that rely too heavily on hindsight. This is especially the case when the explanation for failure essentially comes down to some variant of the 'failure of political will' argument. Our analysis will show that there were indeed special features of PASOK as a political party which made it an unsuitable agent for promoting an alternative economic strategy. To this may be added certain other contingent features of both the Greek economy and society which did not make it fertile ground for such an approach. We discuss such issues at some length in order to delineate what we term the political economy prerequisites for a planning alternative.

But it is important that the analysis does not end here. For the political economy prerequisites often do no more than restate the nature of the problem, in a more revealing way perhaps, but they hardly constitute a full answer to what is entailed in a more successful alternative economic strategy. We argue that PASOK was deeply influenced by the discussions on planning and economic policy which were common to most of the European Left in the late 1970s and early 1980s. It is for this reason that we critically examine this approach in the first half of the book. And for this reason it is also important to consider whether possible inadequacies, gaps or grey areas in the economic policies promoted by the left may have played their part in PASOK's failure. We show that in certain crucial areas this was indeed the case and it is in this sense that we also consider the coherence of the alternative model of economic policy by examining the lessons from the Greek case. There is one final reason to examine the PASOK experiment in the first period (1981-85), although this feature is not discussed at any great length in the book. PASOK's demise and catastrophic final two years can best be understood by analysing its first, and in many ways, formative period in government.

This book is organised in eight chapters. Chapter one sets the scene by examining the relationship of alternative economic strategies to questions of socialist transition. The left social democratic approach is defined as lying somewhere between the revolutionary perspective and that of traditional social democracy as it was experienced in the post-war era. This entails an acceptance of parliamentary democracy and pluralism as the context of any alternative economic strategy. But it also entails a sustained critique of what we term right social democracy. Left social democratic arguments for treating the economy as a system with structural properties, for adopting a political economy perspective and widening the scope of economic policy are examined at some length. Left social democrats would argue that such an approach is necessary if traditional socialist goals for equality, genuine economic pluralism and a dispersal of economic power are to be realisable.

In chapter two we look at the economic theory and practice of indicative planning. It was the perceived failure of indicative planning and macroeconomic policy in the 1960s and 1970s that led many socialist parties in the 1970s to propose more radical policies. In chapter three we discuss the rationale, scope and coherence of the left social democratic alternative. We examine if there exists a coherent package of institutions and mechanisms, which could be said to constitute such an alternative. In chapter 3.8 we present a planning schema which incorporates a range of institutions and mechanisms which have figured in the literature and which have typically been supported by those in the left social democratic tradition. Such a schema, while neither a blue-print nor comprehensive, can act as a useful benchmark for our discussion of the Greek case.

In chapter four we seek to determine that the strategy of

PASOK as it developed up to the 1981 election shares, both in its conceptualisation of the scope of economic policy and in the set of economic policies promoted, many of the characteristics of the left social democratic approach.

In chapter five we examine PASOK's macroeconomic policy up to the June 1985 election, and in particular its strategy of gearing macroeconomic policy to its planning and supply-side strategy. In chapters six and seven we critically assess the supply-side and interventionist aspects of PASOK's alternative. We look at its policy for democratic planning, active planning and industrial strategy, the socialisation of public firms, the role of the new state holding company and so on. Our discussion here constitutes an essential test-case for the institutions and mechanisms of interventionist planning.

Finally in chapter eight we conclude by examining the relevance of the left social democratic model for a country like Greece, and conversely what can be learnt about the model, including the existence of any gaps or inadequacies, from the Greek experiment. Some conclusions of wider relevance are also drawn.

1 Alternative economic strategies: socialism, institutional change and economic policy

The alternative economic strategies which we discuss in this book encompass some conception of a transition to socialism, to a different kind of socio-economic system. Social democracy of course is not only a theory about the transition to socialism but also an ethical stance and a political movement. Given this constant interaction between theory and practice, it is impossible to arrive at the fundamental and underlying theory. However, we attempt to delineate the social democratic path to socialism from what we call the revolutionary path, as well as to distinguish right and left versions of social democracy. For our eventual aim is to examine the alternative economic strategy of the Greek socialist party (PASOK). PASOK defined its strategy as lying between traditional social democracy and the revolutionary path, constituting a 'third road to socialism'. It is, in our typology, a variant of the left social democratic approach.

The relationship between means and ends has bedevilled the socialist movement since its beginnings in the nineteenth century. Certain ethical demands - equality, the abolition of exploitation and the extension of democracy to the economic sphere - have been shared by nearly all strands of thought within the socialist movement. But the precise institutional framework of a future socialist society and the means to ensure the materialisation of such ethical demands has been an area characterised by vagueness, and often, bitter controversy.

The reticence of early Marxist or socialist theorists to be drawn on the 'blue-prints' of a future society, or the nature of the economic relationships and economic policy within such a society, is well-known. As Anderson (1983, pp97-9) has pointed out this vacuum was filled by 'untransformed residues of the tradition of utopian socialism'. But such formulas as 'the replacement of the government of men by the administration of things' or 'the abolition of the division of labour' entailed a severe underestimation of the complexities of any future socialist society. What was lacking was a coherent conception of the institutional specification of any future society.

Clearly any such specification involves more than just the role of economics. It would encompass the 'political structure of socialist democracy', the 'socio-cultural pattern of libertarian levelling' and the 'international relations between - inevitably - unevenly developed socialist countries themselves' (Anderson, 1983, p99). However our main concern is with the 'pattern of an advanced socialist economy'. Anderson usefully lists some of the central questions which have been subject to both a lack of theory and an absence of specific proposals: the range of forms of social ownership, the extent to which the plan could and should replace the market, the ability of planning to adjust to dynamic new needs, the mechanisms for conflict resolution between central and regional interests, the desirable patterns of technology and distribution of labour-times, and the processes to harmonise the interests of consumers with those of producers.

Some of the issues involved have been clarified by such works as Nove's <u>Economics of Feasible Socialism</u> (1983). However the means to arriving at such an economy are no less contentious than the ends themselves. As Anderson again notes, Nove's socialist economy is far more radical than anything achieved by any social democratic government, entailing an abolition of private ownership in the principal means of production and a firm limit to the permitted degree of inequality. Such a project is likely to face in the transition severe opposition. As Anderson (1983, p103) concludes Nove fails to confront such issues:

> ...to construct an economic model of socialism in one advanced country is a legitimate exercise: but to extract it

from any computable relationship with a surrounding, and necessarily opposing, capitalist environment... is to locate it in thin air... In putting that history out of mind, <u>The Economics of Feasible Socialism</u> falls subject to the very criticism it so often makes of Marxism: it proceeds on the basis of manifestly unrealistic assumptions about how people behave - once they are organised in antagonistic classes. In that sense, only a 'Politics of Feasible Socialism' could rescue it from the realm of utopian thought it seeks to escape.

It is with this 'politics' that we are concerned in this chapter by examining the political economy of the left social democratic path to socialism.

In sections 1 and 2, we try to outline what is essential in a left social democratic path to socialism, distinguishing it from both revolutionary politics and right social democracy. In section 3 we seek to find if there is a distinctive left social democratic approach to economic science and policy.

1.1 The revolutionary path to socialism

The transition to socialism has a very precise meaning in the vocabulary of revolutionary socialism. It is not the time-scale of the transition which is of importance - even Stalin (1953) accepted that the transition would need 'an entire historical era' and not 'a fleeting period of 'super-revolutionary' acts and decrees'. Rather it is Lenin's belief that 'the fundamental question of every revolution is the question of power' and that this necessitates a violent revolution to destroy the apparatus of state power of the ruling class.[1] In his famous letter to Kugelman in April 1871, Marx had argued that the 'smashing' of the bureaucratic-military machine 'is the precondition for every real people's revolution on the Continent'.

Here the transition to socialism can be equated with the period of 'building socialism', 'the construction of a new type of state and economic order', <u>after</u> political power has been won (see Anderson, 1980, p194). It is important to note the limits such a conception places on the role of economic policy and theory. It is not surprising that most Marxist economics has focused on such questions as the operation of the capitalist mode of production, the labour theory of value and imperialism.

If the emphasis is on the 'construction of a new type of state and economic order' after the achievement of political power, there is little point in left economists analysing the economic policies a left government will have to introduce in a context of parliamentary democracy and an overwhelmingly market economy. In the revolutionary perspective, economic strategy, before power has been achieved, has little autonomy - it is the political 'moment' which is decisive.

Such an approach is based on a belief that no transition to socialism can be initiated which does not lead to a severe economic crisis. Any attempt by a reforming government to implement a change from one social system to another will lead either to the abandonment of reforms or to a decisive civil war. As a consequence the dominant strategy of revolutionary socialism relies on a conception of dual power: 'Communism denies parliamentarism as a form of society of the future. It denies the possibility of taking over parliament in the long run; sets itself the aim of destroying parliamentarism. Therefore there can only be a question of utilising the bourgeois state institutions for the purpose of their destruction'.[2]

Rather than attempting to exploit the contradictions within the capitalist state, the strategy rests on the creation of organs of socialist democracy, parallel and outside the existing power structures. The model for such a strategy is obviously the experience of Soviets in the Russian revolution. The organs of dual power would then have a decisive role in the political 'moment' of the revolution.

Much of the theory of revolutionary marxism in the first two decades of the century was, in contrast as we shall see to the social democratic approach, aimed at playing down the importance of incipient forms of socialism within capitalism itself (see Parkin, 1979, p147). In parts of his work, notably in Das Kapital, Marx emphasised such new phenomena as the joint-stock company and the growth of workers' co-operatives as constituting embryonic forms of socialism within capitalism. However the revolutionary, and in particular Leninist, case rested, in part, on the argument that there was a basic asymmetry between the bourgeois and proletarian revolution. Lukacs (1971) argued that feudalism had been steadily eroded by capitalist property and market relations. The bourgeoisie did not need a political consciousness to further its own interests. Rather these could be promoted by capitalists merely carrying out their economic function.

However Lukacs argued that for a successful proletarian revolution, the proletariat must be politically conscious of its historic role. This entailed that the proletarian revolution was likely to be more difficult and involve more severe dislocation in the existing socio-economic order (see also, Anderson, 1980, p195). Central to Lukacs' conception was the fact that the short-term economic interests of workers did not immediately coincide with longer-term ones of workers as a class.

As Przeworski (1985, p20) points out:

> Class interest does not necessarily correspond with the interests of each worker, as an individual. Individual workers as well as those of a specific firm or sector have a powerful incentive to pursue their particularist interest at the cost of other workers unless some organisation, a union, a party, or the state has the means to enforce collective discipline. Hence in order to overcome competition, workers must organise as a collective force.

Thus Przeworski notes an increase in wages is in the interest of all workers but a minimum wage, by affecting the relationship among workers, may damage some individual interests.

The importance of this non-coincidence of the short and long-term interests of workers has been emphasised because it has important implications for any transition to socialism. Thus Tawney argued that a socialist party cannot merely present a programme consisting of a shopping list of the demands of its supporters, but must promote a 'creed' to enable it to win and exercise power. Furthermore, this would sometimes conflict with its supporters immediate interests (see Lukes, 1984). As Tawney concludes, a socialist strategy entails: '....not the passage of a series of reforms in the interests of different sections of the working class. It is to abolish all advantages and disabilities which have their source, not in differences of personal quality, but in disparities of wealth, opportunity, social position and economic power. It is in short..... a classless society' (quoted in Lukes, 1984, p272). This, as we shall see, has important implications for the economic theory and policy of social democracy.

We can conclude here by delineating the three major objections to the social democratic approach which constitute the essence of the revolutionary perspective. The response to each is critical to an understanding of the social democratic alternative.

The first aspect is the belief that the very nature of the capitalist mode of production does not allow for a peaceful and unilinear path to socialism. The revolutionary perspective would stress the fact that working class organisation and struggle inherently comes in waves, and/or is often spontaneous. A peaceful and drawn out transition often entails social democratic governments or organisations having to deradicalise their own popular base (see Hall, 1983).

The second aspect is the emphasis on the unity of progression of the capitalist mode of production. This means that reforms, such as welfare expenditure, can only go so far before the confidence of private investors, and thus investment, is affected. What is not possible is to adopt 'salami' tatics to the various branches of investment, for the profitability of one branch will affect all other branches. This implies, as we have seen, an economic crisis during any attempt to institute a socialist transformation (see Przeworski, 1985, p46).

The final aspect is the integral nature of the capitalist state (see Anderson, 1977). As Poulantzas (1973) argued, in particular in his earlier work, the representative institutions of modern democracies have behind them the 'repressive state apparatuses'. In any crisis for the capitalist system as a whole, these, especially the army and the police, can be expected to play a critical role in defense of the existing system.

1.2 Social democracy and socialism

Terms such as social democracy, reformism and revisionism have tended to be both highly ambiguous and deeply controversial. So we must start with some working definitions. For Marx, revolution implied the transition from one social system to another. Reformism, of which one type or another has been the hallmark of social democracy, we take to mean a transition by an accumulation of reforms. The mere management of a social system (revisionism), with some bias towards a particular constituency, does not constitute reformism.

In the early history of the social democratic movement the issue of transition remained obscure. The ambiguity of the

German social democratic party's (hereafter SPD) Erfurt programme of 1891 consisted precisely in the conflict between its immediate programme of objectives and reforms, written by Bernstein, and the more radical implications of its theory and longer-run goals written by Kautsky. While Bernstein's (1909) later revisionism was seen by him as merely bringing the theory of the SPD in line with its day-to-day practice, the bitterness of the ensuing controversy, not only from revolutionaries such as Luxemburg but ultra-orthodox social democrats such as Kautsky, eloquently testifies to the confusion entailed in the social democratic approach.

However some issues were not in contention. Drawing on Engels' later contributions, it was clear that social democrats were committed to winning a majority of the electorate (Colletti, 1972b). Given such theories as the proletarianisation of society, this did not seem particularly problematic, and election results in a number of countries in the early part of the century suggested that the historical tide was very much in social democracy's favour (Przeworski, 1985, p18). Action outside the parliamentary arena was never rejected but gradually become a de facto secondary consideration (Hodgson, 1977c, p121). As Kautsky argued in 1909: 'Social democracy is a revolutionary party, but not a revolution-making party. But we also know that it is no more in our power to make a revolution than it is in the power of our enemies to prevent it. We have no wish either to stir up revolution or to prepare the ground for one' (Quoted in Parkin, 1979, p164). A sharper distinction with the strategy of dual power could hardly be made.

Kautsky's marxism was both deterministic and economistic - it was the development of the forces of production which ensured the victory of socialism (Colletti, 1972b; Laclau and Mouffe, 1981). In the meantime social democrats, while waiting for the eventual electoral majorities, could promote the interests of their supporters by such reforms as better labour legislation, minimum wages and increased welfare expenditure. Furthermore 'it would be a profound error to imagine such reforms could delay social revolution' (Kautsky, quoted in Przeworski, 1979, p30). Thus the essence of reformism was the belief that such reforms were 'steps' on the road towards a socialist transformation of society. Reforms were considered both cumulative and irreversible (Przeworski, 1985, p31).

The coherence of this social democratic approach was first

challenged, at the theoretical level, by Bernstein (1909). Bernstein in an attempt to reunite social democratic theory and practice, based his argument on what he considered significant developments in capitalist economies. He argued against 'breakdown' Marxist theories which suggested the eventual collapse of capitalism under the weight of its own contradictions. He also dismissed theories of the increasing proletarianisation of society and pointed to the continuing importance of the middle classes.

As Colletti (1972b) has forcefully argued the main contradiction for Bernstein is between a democratic political structure and capitalist exploitation. With the increase in the former, the latter would be increasingly untenable - it was now possible to conceive of democratic pressure restricting the worst features of capitalism. Luxemburg (1970) in her classic reply <u>Reform or Revolution</u> pointed out that such an approach lay outside both traditional social democracy and revolutionary theory: 'Revisionism does not want to see the contradictions of capitalism mature [nor to] suppress these contradictions through a revolutionary transformation. Rather it wants to lessen, to attenuate, the capitalist contradictions'.

However revisionism was not based on merely theoretical considerations. Przeworski (1985, chapter 1) has argued that central to the social democratic theory and practice was the decision to participate in elections, to seek extra-working class support and to seek reforms within capitalism rather than dedicating <u>all</u> efforts to its complete transformation. There are two crucial aspects here. Firstly, that in the inter-war years social democratic governments came to power as minority governments or lacked a full mandate for the implementation of socialist long-term goals. Secondly, as we have indicated, socialist theory was not very clear on the nature of a future socialist economy: 'The choice of industries which were to be nationalised, methods of financing, techniques of management and the mutual relations among sectors turned out to be technical problems for which social democrats were unprepared' (Przeworski, 1985, p34). In short social democrats lacked an economic policy of their own.

As we shall argue in section 3, this was rectified by the marriage of social democracy to Keynesianism. It was this which provided an economic theory for Bernstein's attentuation of capitalism. It now seemed possible to promote a strategy of redistribution, welfare reforms and full-employment without transforming property relations and by using indirect means to 'control' the economy. In contradiction to revolutionary theory that gave little scope for economic policy before the revolution, here there was not only space for such an economic policy but it also came to be seen as an alternative for any revolution. But this entails the abandonment of reformism which had always meant that cumulative reforms would lead to a transformation of the social system. It is this general approach that we term right social democracy. In particular the leadership of most socialist and social democratic parties since the war have been fully committed to this abandonment of reformism, even if all have had significant left social democratic opposition within their parties.

If Bernstein is the theoretical antecedent of this approach, then the school of Austro-Marxism made an important contribution to what we shall call the left social democratic approach.[3] The Austro-Marxists sought explicitly to distinguish themselves from both the abandonment of reformism by social democracy and the revolutionary perspective of the Third International. Once more the analysis was based on an interpretation of developments within capitalism.

Hilferding's concept of 'organised capitalism' was based on his study of imperialism and finance-capital. In contrast to Bernstein he argued that the centralisation of capital, the growth of monopoly and cartels, and the rise of finance capital could not prevent economic crises. But unlike Lenin, who tended to conceive imperialism as the final stage of a moribund capitalism, Hilferding argued that such developments created the preconditions for a future rational organisation of the economy, in which the state was likely to play a major role. The state 'begins to assume the character of a consciously rational structuring of society in the interests of all; it establishes the organisational preconditions for socialism' (Bottomore and Goode, 1978, p24). Another Austro-Marxist, Renner, argued that the 'laissez-faire' model of capitalism was no longer of any relevance, and that the state could be used for socialist ends.[4]

Otto Bauer's political strategy of the 'slow revolution' is thus a predecessor of left social democracy. It was a strategy of 'revolution through reform' which sought to link the opposition gains achieved by social democracy with the longer-term objective of replacing the capitalist system altogether: 'the conquest of power by the working class had to be accompanied by a gradual, patient construction of social institutions' (Bottomore and Goode, 1978, p39). Here too there is space for the autonomy of economic policy. If the task is one of 'slow revolution', then appropriate policies for planning or mediating between the short and long-term interests of workers become of paramount importance. The importance given to institutional changes and their relationship to transforming the nature of economic power within society, is, as we shall see, an integral part of the left social democratic approach.

A further critical distinction between left and right social democracy is their respective approach to the questions of democracy and the state. In the 'classical' era, most parties of the Second International saw themselves as extending democracy from the political to the economic and social spheres. However things can be put a little more concretely. Social democracy did not only accept democracy in general, but parliamentary democracy. Thus the Eurocommunist movement in the 1970s, in part to distance itself from the Soviet model, made it clear how strong a break from the revolutionary tradition had been achieved, by declaring its commitment to parliamentary democracy in both the transition to, and arrival at socialism (Carrillo, 1977).

However the Eurocommunist strategy, which is one example of the democratic road to socialism, also sheds light on some of the existing different approaches to both the state and democracy. As Jessop (1982, p14) has argued:

> Rightwing Eurocommunists tend to view the transition as gradual and progressive, based on an anti-monopoly class alliance under the leadership of the Communist vanguard party, and orientated to the strengthening of parliamentary control over the state and economic systems in association with certain measures of trade union participation and economic planning. Left Eurocommunists tend to view the transition as a long series of ruptures or breaks, based on a national popular, broad democratic alliance involving new social movements as well as class forces and organised in a pluralistic manner, and orientated to the restructuring of the state and economy so that there is extensive democracy at the base as well as an overarching, unifying parliamentary forum.

Thus while any left social democratic strategy necessarily rejects any dual power conception, there is considerable scope for disagreement on the extent to which the state itself must be restructured, and the extent to which forms of representative democracy can be articulated, as Poulantzas (1980a) envisaged, with more direct and participatory forms.

Finally, left social democracy has also been more aware of the fact that since its strategy challenges the existing social system then it may face opposition that threatens democratic institutions.[5] Bauer's notion of 'defensive violence' is a characteristic response. In 1926 the Austrian social democrats at Linz outlined the conditions under which violence may be necessary to protect civil, political and social rights which have been achieved by democratic means.[6] More recently Rowthorn (1981) argued for the need for such issues to be discussed in the context of British left debate on the Alternative Economic Strategy.

We have attempted here to outline the main features of the social democratic approach to socialism, as well as drawing a distinction between left and right social democracy. Obviously our rather schematic account cannot hope to do justice to the wide variety of approaches within the socialist tradition. Few parties or movements will neatly fall into any one of our categories, since most incorporate a wide range of views and any programme is open to an infinite number of interpretations as to its 'true' logic. Furthermore the distinction between left and right social democratic strategies is not easily made in practice. For as Esping-Anderson (1984) has pointed out the issue of distinguishing reforms within capitalism and reforms attempting its transformation is complicated by the fact that it may not be clear when a given policy has long-term revolutionary consequences. However the differences between left and right social democracy are further developed in the following section when we examine their divergent views with respect to economic theory and policy.

1.3 The economics of social democracy

The argument presented here is that right social democracy can be distinguished by its association with the post-war Keynesian consensus and its acceptance of orthodox neoclassical economics, as this was understood in the post-war Keynesian-neoclassical synthesis. Left social democracy, on the other hand, can be distinguished by its critique of such an approach and its support for a political economy perspective. This distinction will provide the context for left social democracy's rejection of an economic policy primarily based on Keynesian demand management and indicative planning (chapter 2) and its conception of an alternative, and more interventionist, economic strategy (chapter 3).

1.3.1 Economics and the social democratic consensus

To understand the nature of the social democratic consensus we must look at the structure of neoclassical economics, the Keynesian 'revolution' and the implicit political compromise on which the consensus was founded. For as Steindl (1985, p103) has argued, in order to understand 'the relations of knowledge and material development of history we must free ourselves of primitive ideas of unilateral causation ...the idea or knowledge must combine with other events or developments in order to become relevant ...the idea (consciousness) exists between a flow of history which produces it and another flow of history which receives it as a seed'.

The neoclassical paradigm, which entails a particular conception of the relationship between economics and society, has dominated economic theory in this century. Here we can sketch some of its salient characteristics which are important for our purposes. The specification that the universal economic problem is one of scarcity means that the main task of theory is to examine the allocation of scarce resources to unlimited ends. This limits the scope of economic science, and, as we shall see, presents an ahistorical view of the world.

The basic building-block of the paradigm is the rational self-interested individual, who must work out the optimum response to changing price signals. The emphasis is almost exclusively on exchange relations for: '...given the fact that society is seen as an agglomeration of individuals whose nature is fixed, who do not combine together in a social production process and whose only link with each other is through the buying and selling of market commodities, market phenomena must inevitably assume primary importance' (Rowthorn, 1980, p16). The basically long-run concept of equilibrium and the study of the existence, uniqueness and stability of such an equilibrium have dominated the pure theory of neoclassical economics.[7] The work of such theorists as Arrow, Hahn and Debreu has attempted to demonstrate rigourously the existence of such an equilibrium, although the question of stability still constitutes a major problem in neoclassical economics (see Arrow and Hahn, 1954; 1971; Debreu 1959).

The emphasis on market relations and equilibrium, while not logically entailing, does suggest a perspective of social harmony in which any conflict is in principle reconcilable. Recurrent phenomena such as unemployment or economic crisis can be regarded as deviations from equilibrium (due to 'frictions' or 'imperfections') and as neither intrinsic nor functional to the capitalist economy (Rowthorn, 1980, p18). As Bleaney (1985, p6) points out:

> the notion of general equilibrium immediately suggested the very static idea that if only all prices were at their equilibrium values, macroeconomic bliss would automatically follow. So macroeconomic disequilibria of any sort could be regarded as a problem of some prices being out of line with others, and the solution would lie in relative price movements.

From this we can see how the most common response in the 1920s to unemployment emanating from neoclassical economics was to lower wages. Given the goals of social democracy this was not a very promising prospect. As we have seen, in this era, lacking a mandate for nationalisation on a large scale, left social democrats were left without a distinctive economic strategy of their own. Thus to understand the eventual 'marriage' of social democracy to orthodox economics, we have to examine developments within orthodox economic theory. As Durbin (1984, 1985) has argued, a number of Labour economic thinkers in Britain were to build on Keynesian ideas and the new microeconomics of imperfect competition to create a case for socialism on, what she has termed, the 'New Socialist Economics'.[8]

The 'new microeconomics' of imperfect competition associated with Joan Robinson (1933) and Chamberlain (1933) opened the way to policies of the public perfection of markets, based on market failures. One such market failure was the existence of monopoly - in the US, Berle and Means had pointed to the extent to which the economy was dominated by huge corporations (Sweezy, 1972a). By now there is a vast literature on a wide number of market failures and the rationale for state intervention (Stiglitz, 1986). As Murray (1984, p215) has argued this approach entailed a particular conception of intervention:

> ...it is not the market as a mechanism which is held to be at fault. Merely that its particular signals are not accurate, and that state intervention is needed to correct them; taxes, subsidies, a wealth tax, anti-monopoly legislation, infant industries tariffs and so on. The market remains the dominant economic nexus. Modifications to it can be made at the level of circulation without reference to production.

Of equal importance to the 'New Socialist Economics' was the Keynesian revolution, although the nature of that revolution has been highly contested. It is common ground that Keynes (1964) held that the price mechanism alone could not be relied upon for the problems of stability in a capitalist economy - more controversial is precisely why Keynes thought this should be the case. Keynes was particularly concerned with the fact that in an unemployment context reducing wages and prices could actually accentuate the problem because of its effect on business expectations about future effective demand and the tendency of disturbances in one market being transmitted to another. Fisher (1933) in the US, who was influenced by the effect of bankruptcies, argued on similar grounds (see Tobin, 1980). Keynes' work seemed to have removed the stranglehold of Say's Law and provided a rationale for government intervention.

However as Schott (1982) has argued Keynesianism was never a single entity, nor did it have any unambiguous policy implications. The dominant interpretation in the post-war era came to be known as the Keynesian-neoclassical synthesis. This implied that Keynes' significance was more at the level of policy, the need for demand management, than at the level of economic theory (Bleaney, 1985, p22). Modigliani (1944), for instance, suggested that Keynes' contribution could be seen as a special case of neoclassical theory, when either wages were fixed or there was a liquidity trap. Modigliani had reached such a conclusion by asking whether it is possible to have a long-term unemployment equilibrium - once more indicating the stress neoclassical theorists place on looking at long-term equilibrium. This placed less emphasis on Keynes' insights into the nature of the investment function and expectations and the role of money and financial institutions.[9]

Such an approach had important consequences for the scope of economic policy, for:

> the management of the level of aggregate demand is controversial only in terms of the relative sequence and speeds of adjustment of the various markets and economic variables. In this light, different policy instruments concerned with, for example, the rate of interest, government expenditure and the budget deficit, the money supply and taxation are seen as effective according to their effect on aggregate demand and subsequent output and employment. Debate can be confined to these considerations, and the wider social conditions of state intervention can be safely left to the microeconomics of externalities, public goods and natural monopolies, or to the ethics of normative economics. (Fine, 1980, p51).

We would argue that this is the dominant way that social democracy has been experienced as government. The essence of the approach was 'intervention at a distance', implying an abandonment of more interventionist means and a considerable reliance on the market itself as an instrument of control (Thompson, 1984; Goldthorpe, 1985). In Britain direct controls in the economy were quickly replaced in the post-war era by the 'Butskellite' consensus of fine-tuning demand management. When planning was thought appropriate it was of the indicative variety (see chapter two). Right social democracy shared the confidence of the economics profession in the 1950s and 1960s. Samuelson's well-known remark that 'as we no longer meekly accept disease, we no longer need accept mass unemployment' is indicative of this confidence. In Britain Crosland (1956, 1974) argued that there was now no insurmountable problem to controlling the capitalist economy. Fabians in the post-war era argued that the mixed economy was a permanent resting place, and that the gains of Keynesianism and full-employment would never be challenged by the Conservative opposition (Durbin, 1984, pp47-9). This was a reaffirmation of the belief in the irreversibility of social democratic reforms. This approach is also at the heart of the SPD's 'Bad Godesberg' programme (1959), which is perhaps the most coherent statement of right social democracy.

For right social democracy there were two further consequences. Firstly right social democrats no longer sought to differentiate themselves on the terrain of economic policy. While they may pride themselves on their greater capacity or willingness to use the new methods of controlling the economy, economic problems were basically technical ones. The means of economic policy were not different in kind to those that might be employed by any government. Thus, as Durbin (1984) has argued, for many social democrats, economics was about efficiency and socialist politics could be reformulated on the terrain of ethics and justice, by such goals as income redistribution or expanding the welfare state.[10] Here we can detect an acceptance of economics as a value-free science, in terms of Robbins' classic distinction between positive and normative economics: 'The myth of a value-free economics depended on a political consensus as to the economic objectives. The post-war era was marked by widespread commitment to full employment and economic growth. Orthodox economics took such shared objectives for granted and discussed the technical means of their achievement' (Hodgson, 1984, p19).

Secondly, it was now possible to promote socialist goals without a major challenge to the institution of private capital: 'social democrats defined their role as that of modifying the play of market forces, in effect abandoning the project of nationalisation altogether' (Przeworski, 1985, p37). For theorists such as Crosland, the separation of ownership and control in modern capitalism meant that private property was no longer the central question for socialists. As Keynes (1964, p378) himself had written, 'it is not the ownership of the instruments of production which is important for the state to assume. If the state is able to determine the aggregate amount of resources devoted to augmenting the instruments and basic reward to those who own them it will have accomplished all that is necessary'.

Keynesianism and the economics of market failures not only gave social democrats a role in government they also fitted well with their political values. As Przeworski (1985) has argued, orthodoxy had argued that unemployment could be cured by reducing wages - wages were seen as a 'particularist' interest, cutting into profits and investment. The interests of the capitalist were universalizeable, reflecting the interest of society as a whole. Keynes' argument on the problem of effective demand, on the other hand, suggested that taxation could be used to redistribute income and expand state expenditure. If lower-income families had a higher propensity to consume, 'the significance of increasing wages changed from being an impediment to national economic development to its stimulus' (Przeworski, 1985, p37). This in part also explains the importance of the welfare state - for now it was workers, or, more generally, the people who were the bearers of the universal interest.

Thus to see the full significance of the 'marriage' of Keynesianism and social democracy we must examine the implicit political compromise that sustained the consensus in the post-war era. Przeworski (1985, p207) has described this as a compromise between democracy and capitalism:

> those who do not own the instruments of production consent to the private ownership of the capital stock while those who do own the productive instruments consent to political institutions that permit other groups to effectively press their claim to the allocation of resources and the distribution of output... It was Keynesianism that provided the ideological foundations for the compromise of capitalist democracy.

Thus if the compromise entailed social democrats abandoning their challenge to capitalist property rights and limiting their economic intervention to indirect means, it also entailed a commitment to higher wages, full-employment and the welfare state.

It is by now commonplace to point out that it was precisely this social democratic consensus that was challenged by the new economic conditions of the 1970s. There are three closely interrelated aspects to this phenomenon. The first was the end to the long post-war boom in the capitalist economies, and the new economic context of stagflation. The reasons for this go beyond our scope here but they include the rise of working class power after two decades of relatively full employment, the decline of deferential behaviour, the fiscal crisis of the state, the decline in the economic hegemony of the US registered by the withdrawal of the dollar as the reserve currency and the OPEC crisis (Goldthorpe and Hirsch, 1978; Marglin and Shor, 1990).

Secondly there was a decline in the efficacy of traditional Keynesian demand-management measures. At one level this reflected an increasing emphasis on the supply-side problems of the economy that it was felt that Keynesianism, with its concerns over the short run and the gap between actual and potential output, had neglected (Przeworski, 1985, p211). However it also reflected the political economy conditions which are a prerequisite for Keynesianism. Bleaney (1985) argues that the social democratic consensus built up high expectations of wages earners in respect to productivity and if these are not removed the expansion of the economy 'will hit an inflationary barrier on industrial militancy as economic prospects improve'. An interesting part of his argument is that 'crowding out' is not a different argument against Keynesian prescriptions. In the 1930s governments could use an accomodating monetary policy to deal with the interest rate. But the restrictive monetary policies of the 1970s and 1980s:

> represented a well-publicised signal from governments that employment would not be insulated against the effects of excessive wage increases... But having taken on this role of an informal incomes policy, monetary targets cannot be released from it until the threat of a wage explosion has been completely removed... Whilst macroeconomic policy is designed to resolve this difficulty, it cannot be switched over to the conventional objective of controlling the level of output. (Bleaney, 1985, p204).

The third aspect of the decline of the consensus was the rise of monetarism which 'was just one aspect of a forceful challenge to the post-war Keynesian consensus... It was not simply a debate over the technical causes of inflation. The New Right was asserting a conception of what sort of economic system was desirable...' (Hodgson, 1984, p19). It is difficult to accept that the dominance of monetarism is due to its theoretical superiority over Keynesian economics. The monetarist prescriptions come from models in which prices are flexible - in the extreme case of New Classical theory there is instantaneous market clearing in all markets at all times. As Hahn (1984) has argued in a world adequately described by some version of Walrasian equilibrium, Keynesian policy is not so much wrong, as nonsensical - there is literally no need for it.

That is not to say that monetarist economic strategy had no rational kernel. On the one hand, it is certainly the case that actual monetarist experiments in the 1980s (notably in Britain and the US), contrary to the predictions of their theoretical models, created a recession and a large increase in unemployment. On the other hand, it would be misleading to think that monetarism was concerned only with the reduction of inflation. For in actual fact monetarism both as a theory and when implemented was tied to a number of supply-side policies. It sought to weaken labour, either directly through unemployment or because workers knew that if their firms went bankrupt due to high wage claims, the government would not intervene. The strategy also entailed such measures as privatisation and deregulation to widen the scope of the market mechanism. Its budgetary policy was an attempt to limit the extent to which democratic pressure could increase public spending for collective needs. Implicit in the whole strategy was the idea that Britain's economic problems can be traced to certain structural problems which have fettered entrepreneurship and capitalist profitability. Implicit in the monetarist strategy was a shift in the balance of power towards the private sector (see Gilhespy et al, 1986, p35; Murray, 1984).

Thus the decline of the social democratic consensus led to a radical response: 'that is for proclamations of the 'death' of Keynesianism and for the development of alternative conceptions of economic policy, entailing far reaching reassessments of the proper ends of such policy as well as of its means... these new developments reveal departures from the post-war consensus which go in quite different, indeed contrary, directions in the role they would accord to government - and hence politics - within the economic sphere' (Goldthorpe, 1985). If monetarism challenged the means and ends of economic policy of the consensus from the right, left social democracy was to carry out a similar exercise from the left.

1.3.2 Left social democracy and political economy

The argument in this section is that left social democracy can be distinguished from right social democracy by its support for a wider conception of the role of economic policy. Although it is clearly a simplification, we argue that, just as right social democracy's more limited conception of the role of economic policy is tied to its adoption of orthodox economics, so left social democracy's stance is related to its adoption of what we call the political economy tradition. Of course, in actual fact there is no one political economy tradition. Rather the term political economy is used here as a convenient shorthand for a number of schools of thought. Prominent amongst these is the Marxist tradition, that of Post-Keynesianism which has drawn on the work of Kalecki and some features of Keynes, and a number of groups which have come to political economy from political science rather than economics.

There are three interrelated themes in this section. The first is that left social democracy needs a realistic understanding of the capitalist economy and its historical development. Secondly, that the nature of orthodox economic theory is such that key features of the capitalist economy are either disguised or not treated at all. In particular, relations of economic power and the role of institutions critical to the workings of the capitalist economy are rarely integrated into the analysis. The third theme is that left social democracy does not only adopt a political economy view in order to understand how the capitalist economy works, but also to investigate the extent to which the structure of the capitalist economy places severe limits on the implementation of its goals.[11] It is this interrelationship between a political economy understanding of the capitalist economy and the goals of left social democracy which is the basis for the wider scope of economic policy supported by left social democracy.

Lange (1935) argued that the market model can exist under various socio-economic frameworks - the capitalist economy is not the same as the market economy and to the extent that orthodox economics is merely a theory of distribution of scarce resources between different uses, then it does not need any sociological data and is not really a social science at all. Lange was particularly concerned with defending Marxist economics, and by its very nature Marxism offers a political economy perspective in trying to analyse the social relations which underlie the economy, the conflict involved in the production and distribution of the social product and the role of private property. But, his position would be shared by most in the political economy tradition: 'Marxian economics is distinguished by making the specification of this additional institutional [sociological] datum the very cornerstone of its analysis, thus discovering the clue to the peculiarity of the capitalist system which it differs from other forms of exchange economy'.

Political economists would argue that it is precisely this lack of 'institutional data' which prevented orthodox economics from either predicting or fully understanding the decline of the social democratic consensus, and at the level of theory the decline of the Keynesian-neoclassical synthesis which we described in the previous section:

> to the extent that they [orthodox economists - Goldthorpe refers to the work of Hicks, Solow and Scott] introduced <u>ad hoc</u> such notions as 'a growing worker militancy' or 'codes of good behaviour enforced by social pressures' they were in effect accepting that the reigning paradigms of economic analysis were in themselves inadequate to the task before them. But if, on the other hand, they sought to preserve the intellectual autarky of economics and to reaffirm the basic validity of conventional analyses, then, it seemed they had to build into their explorations very large 'residual categories' (Goldthorpe, 1984b, p2).

As we shall see, the political economy tradition has attempted to integrate political and economic analysis to avoid such large 'residual categories'.[12]

But it is not only with respect to understanding the historic development of capitalist economies that orthodox economics is felt to be inadequate. By examining in more detail three aspects of orthodox economics (namely welfare economics, the doctrine of consumer sovereignty and the lack of institutional analysis) we can shed more light on how it is felt that the structure of orthodox economics disguises important features of the capitalist economy which could obstruct the implementation of any left social democratic strategy.

Recall the distinction made by right social democracy between economics as the science of efficiency and ethics as the praxis of socialist politics. As Edgley (1982) has pointed out:

> As far as 'pure logic' is concerned, to say nothing of actual particular examples, the separation of science as theory from practice as morality, together with the paradigmatic rationality of science, leaves morality a complete free hand, so to speak, to recommend anything whatsoever, revolution included. But such indeterminancy is impossible to live with and social need develops a rationale, for eliminating all but a narrow range of these 'logical possibilities'.

Thus to understand the political economy perspective of left social democracy we must begin by examining the ways it is felt orthodox economics, and capitalism as an economic system, limit the range of these possibilities.

Within orthodox economics, normative questions are the domain of welfare economics. Consider the two fundamental theorems of welfare economics - every competitive equilibrium is Pareto optimal and every Pareto optimal point can be arrived at by a competitive equilibrium. Pareto optimality is distinguished from social optimum which depends on some form of general welfare function. This may suggest that 'politicians' can choose their desired welfare function to add to the economics of efficiency.

However the issue is not so simple. For the theory indicates that lump-sum taxes are the preferred measure for redistribution in that they do not affect marginal utility and cost curves. The problem with this conception is the doubt that exists about the existence in practice of any large amount of reserves of lump-sum taxes in a dynamic economy. And those who have faced lump-sum taxes (for instance British banks facing windfall taxes in the 1980s) have argued that such measures affect their future expectations and thus their actions, considerably reducing the efficacy of lump-sum taxes. Furthermore Pareto optimality cannot incorporate into the analysis inter-distributional questions if this leads to losses of utility for some individuals or groups. Policies which harm certain groups, for instance by restricting income that can be earned from ownership of the means of production can never be considered using the Pareto principle. The range of possibilities is not as large as might appear at first sight.

The approach of welfare economics can be usefully compared to Rawls' liberal theory of justice. One of the most well-known features of Rawls' theory is his presumption for equality if this leads to the greatest benefit of the least advantaged. This suggests that redistribution can be promoted until this leads to 'inefficiency', in the sense that it can lead to a deterioration in the position of the 'least advantaged'. As Levine (1984) has argued this conception is as ahistorical as neoclassical economics. The problem is in stipulating the 'rules of the game': '...Rawls would have us look only at payoffs under different distributions and not put the 'rules of the game' in question' (Levine, 1984, p81). The interesting question is not merely to compare various levels of distribution of income and efficiency with the given institutional arrangements of society but to compare whether, for instance, more radical redistribution is possible once the structure of ownership or other features of the institutional framework are altered.

In reviewing two recent contributions to restating the socialist case in term of values (Crick, 1984; Plant, 1984), Beetham (1985) has clarified left social democracy's rejection of a socialist strategy articulated predominantly on the level of ethics. It is worth quoting at some length:

> I remain unconvinced that the Crosland tradition of defining socialism simply in terms of a commitment to certain values will do. Socialists also have a theory about the structure of power and privilege which prevent these values from being realised... Concentrations of private wealth... also perform as capital, a crucial function in the system of production, and precisely constitute a formidable social power. That function and that power set clear limits to policies for redistribution of wealth and income, they frustrate attempts at the collective control of production, both at the workplace and in society at large.

It is the ahistoricity, or what Lange (1935) has called the lack of institutional or sociological data, of welfare economics, or Rawlsian justice, that prevents their incorporating questions of economic power into the analysis. For left social democrats, Beetham's 'formidable social power' can exercise a real veto over, say, redistribution, given its control over investment decisions which are in turn crucial to the health of the economy: 'the efficacy of social democrats - as of any other party - in regulating the economy and mitigating the social effects depends on the profitability of the private sector and the willingness of capitalists to cooperate' (Przeworski, 1985, p42).[13]

Orthodox economics' inability to confront such questions of economic power is also related to its theoretical stance of methodological individualism and the doctrine of consumer sovereignty. Consumer sovereignty is based on a harmonious picture of society where firms merely respond to the maximising behaviour of individuals. The power of such individuals to consume depends on their income and wealth: 'Exactly who is poor and who is rich is principally a question of economic power over resources... Under the market economic theory such issues are ignored' (Gilhespy et al, 1986, p13). Furthermore the conception of market economics that economic power is adequately distributed among millions of individuals ignores the fact that many central decisions are taken not by individuals but by large institutions, such as companies or banks, and there are no grounds for thinking that their interests automatically coincide with either society's interest or that of consumers as individuals.

It is also clear that firms in capitalist economies do not behave in the passive manner suggested by the consumer sovereignty approach, but actively intervene, for example, through advertising, to shape consumer preferences. Hodgson (1984) uses the work of Best (1982) on the US motor car, tyre and oil companies to show that such behaviour by firms goes beyond advertising to affect consumer choice. The road 'lobby' systematically set about buying and dismantling rail and tram lines in order to promote the market for cars:

> The rail alternative was deliberately destroyed and removed from the competitive process, so that it was no longer possible for any 'individual preference' for public transport to express itself in market terms. From the individual point of view it became 'rational' to use the motor car. What was socially rational was a different matter (Hodgson, 1984, p182).

Such arguments of economic power are seen to counter the doctrine of consumer sovereignty and the presumption against intervention in the market system of traditional microeconomics: 'if we allow firms the freedom to persuade, then we cannot easily disallow government the means to constrain the individual in her consumption patterns. If the consumer is not sovereign then sovereignty is not taken away from the individual by collective action' (Cowling and Sugden, 1987, p127).

An individualistic bias is also in evidence in the Keynesian-neoclassical synthesis. In contrast the post-Keynesian interpretation of the nature of the Keynesian revolution has placed more emphasis on Keynes' discussion of the institutional developments in investment markets, which played an important part in his theory of the rate of interest and the marginal efficiency of capital. In addition, the structure of financial markets and institutions and their propensity to generate instability are seen as crucial to an understanding of the workings of the capitalist economy (see, for example, Minsky,

1982). Unemployment is seen as a more properly macroeconomic phenomenon, not reducable to the aggregate of individual propensities or random shocks. The emphasis is on the intrinsic instability of investment in capitalist economies and the consequent need for more radical measures for the socialisation of investment (Fine, 1980, p51).[14] The importance of these issues can be seen by comparing the approach of the Keynesian-neoclassical synthesis with that of Post-Keynesians, for example. Eichner (1978) has outlined four broad elements in this perspective. Firstly it offers an explanation of growth and income distribution and their inter-relationship. Secondly the economic system is seen as being constantly in motion, challenging the equilibrium notions of orthodox economics. Thirdly emphasis is placed on credit and other monetary institutions which play a fundamental role in the dynamic process being analysed. Fourthly, more realism is placed at the heart of the models, incorporating the role of monopolies, multi-national companies, administered prices, trade unions and so on.[15]

To recap therefore, there are a number of features of political economy which are particularly relevant for left social democracy. Rather than beginning with an agglomeration of individuals maximising with given constraints, the approach seeks to understand the historical specificity of capitalism as a system with structural properties. It therefore seeks more realistic models of capitalist economies, while placing emphasis on institutions and the relations of economic power[16]. As a consequence the political economy approach seeks to integrate political and economic analysis. This has been resisted by orthodox economists. Hahn's (1985) response is particularly representative: 'As a theoretical economist I share the distrust of grand theorising and in fact believe it cannot be done... when the political economist is not being purely descriptive, he provides murky, often barely comprehensible theories which cannot be used to answer any of the important questions of the subject'. Such a formulation, however, prejudges what exactly the important questions of the subject are and mistakes methodology with the object of study. Marx's own work, and the political economy tradition in general, is full of theoretical abstractions. The question is more to specify the important variables and exogenous factors in any theoretical models (see Marglin, 1990).

The above is particularly significant for left social democracy. For it entails that the structural properties of the existing system must be analysed with a view to examining what transformations of the system are required if socialist goals are to be attainable. What is rejected is the right social democratic contention that indirect controls and the market, together with measures to mitigate the inequalities that are inherent in the actual structure of market power and advantage are sufficient.[17] As Przeworski (1985, p41) has argued: 'mitigation does not become transformation: indeed, without transformation the need to mitigate becomes eternal. Social democrats find themselves in the situation which Marx attributed to Louis Bonaparte: their policies seem contradictory since they are forced at the same time to strengthen the productive power of capital and to counteract its effects'.[18] In seeking to transform the existing socioeconomic system left social democracy has not abandoned reformism in its traditional sense. It is for this reason that left social democracy has a wider conception of the role of economic policy.

We can end here with a brief discussion of what this wider conception entails. One thing it entails is a more interventionist economic policy and left social democracy is clearly associated with a more interventionist stance. We have seen in this section a number of reasons why this should be the case. But a 'more interventionist stance' is a rather vague and ambiguous formulation and matters are more complex.

A central argument for a more interventionist stance has always been the role of the 'developmental state'. As Leys (1985, p24) has argued:

> the market, by definition, only registers existing weaknesses and has itself no inherent capacity to convert weaknesses into strengths. It is increasingly clear that the most successful recent industrialising countries owe their success not to economic liberalisation but to active 'developmental' state apparatuses which have identified productive sectors for long-term growth and secured not only sustained high levels of investment in these sectors but also consistent complementary policies across a wide range of fields, from infrastructural provision to fiscal policy, state purchasing and macroeconomic management.

Furthermore left social democratic theorists have argued that the state intervenes not only at the level of circulation, but also at the level of production. Thus right social democracy's concern with Keynesian and other indirect forms of intervention is seen as one-sided (Harris and Fine, 1976).[19] As Eatwell and Green (1984, p202) conclude:

> the market mechanism cannot be relied upon to determine the scale and pattern of accumulation appropriate to the circumstances of the economy, even if full-employment is attainable by suitable intervention. Accumulation involves a process of continual structural transformation, a qualitative change, in which the scale, content and location of economic activity are progressively transformed. There is no automatic mechanism ensuring that this takes place in a desirable manner or at a desirable pace.[20]

Most left social democratic theorists would concur with the 'developmental state' idea. However, the issue is much more complicated than merely determining the extent of state intervention compared to the scope that should be allowed to the market. To understand why this is the case, it is necessary to see that markets are themselves institutions. The argument is that markets do not exist in isolation from the rest of society but through a whole web of institutions (Hodgson, 1986, 1988; Hall 1986).

This conception can be usefully contrasted to that of orthodox neoclassical economics. Here the market is often set up in its idealised perfect competition form and contrasted to the real world where 'imperfections' may provide for inflexibility in the market mechanism: trade union power, restrictions on hiring and firing, the welfare state etc. Hodgson (1986; 1988, especially chapter 8) discusses the alternative approach in which such imperfections, and monopolistic tendencies in the product market or trade unions actually promote the working of the market, in part by reducing uncertainty and instability. In this light the market mechanism works <u>because</u> of these imperfections not in spite of them.

Much of this analysis stems from a reassessment of the importance of information and the role of institutions as providers of information. In the neoclassical paradigm information is costless and abundant - it is this which ensures that all gains from trade are fully exploited. But as Boland has pointed out: 'one of the roles that institutions play is to create knowledge and information for the individual decision maker. In particular, institutions provide social knowledge which may be needed for <u>interaction</u> with other individual decision makers' (Quoted in Hodgson, 1986, p157).

A useful distinction can be made between centralised and decentralised markets in this respect and here the role of transactions costs is particularly significant (Grahl and Teague, 1990, pp45-52). In decentralised markets there may be a lack of a homogenous good, it may be costly for potential sellers and buyers to find each other, and traders may enter into long-term relationships. A good example of such a decentralised market would be the labour market.[21] The implication of this analysis is that:

> ...the efficiency of market exchange cannot be assured by the market itself - some external social support is needed to make an exchange system work. In the case of centralised markets this support is provided by a clearly visible institution: an 'exchange', a clearing house of a financial centre with its rules of procedure and specialist intermediaries. But decentralised, limited information markets also depend on external elements: networks of associated producers which can distribute information on prices and costs; regulatory bodies, private and public, which can reduce uncertainty about the nature of the product to be sold and the quality which should be expected. (Grahl and Teague, 1990, p47).

This conception of the importance of institutions, and that certain markets may actually depend on institutional rigidities, does not of course imply that existing institutions are in any was optimal or beyond reform. Quite the contrary.[22] Indeed there has been much recent research on the institutional determinants of economic performance (see Henley and Tsakalotos, forthcoming). But it does imply that no easy distinction can be made between the scope for the market on the one hand and the role for intervention/institutions on the other.

There are thus two major implications of the above discussion. The first is that the question is not one of intervention versus non-intervention but 'which type of intervention is to be carried out, and for which ends' (Hodgson, 1986, p166).[23] As we shall see in chapter three the forms of intervention promoted by left social democracy entail neither centralised planning nor the complete replacement of the market and private property. The second implication is to go beyond a sterile debate at the theoretical level between the plan and the market:

> ...institutional intervention at both the 'macro' and 'micro' level is required. This could include such measures as structural and/or ownership changes in the banking system and industry, especially those designed to increase participation and accountability, as well as efficiency, in the economy. Intervention on this scale is likely to take place in the context of national planning which is partly indicative, partly regulatory and partly directive in character. The task is to apply the whole literature on, and experience of, planning, industrial organisation, financial structures, and the management of production in a way that is appropriate for economic and social objectives. Commitment to a sole regulator such as 'the market' or 'national planning' or 'effective demand' is neither feasible nor effective. (Hodgson, 1986, p160-1).

Such a conception broadens both the scope and means of economic policy and is clearly distinguishable from a right social democratic perspective. Thus structural or ownership changes in industry, or intervention in the financial or credit institutions, are just as much a part of economic policy as macroeconomic control and demand management. The scope and means of left social democracy are discussed further in chapter three.

However one final conclusion is in order here. Any left social democratic strategy will not be merely an 'economic' question - it is not a matter of developing new technical solutions to technical problems. Any strategy which seeks in some way to transform the existing structure of the socio-economic system and which challenges both the means and ends of economic policy, will necessarily be opposed by those who have most to gain from the existing system and structure of economic power. The extent to which any strategy can generate support and withstand opposition is likely to be crucial to its overall coherence. Thus any left social democratic strategy entails a political economy project.

Notes

1. See Lane (1981) and Harding (1983) for the development of Lenin's concept of revolution.

2. This is the sixth thesis from the Theses on the Communist Parties and Parliamentarianism, accepted by the Second Congress of the Communist International in August, 1920.

3. For a discussion of Austro-Marxism, including some of the central texts, see Bottomore and Goode (1978).

4. The continuity in thinking between such a conception and more recent socialist strategies is striking. Thus the Communist Party of France (hereafter PCF) support for the 'Common Programme' of the left in France in the 1970s was based on its 'state monopoly capitalism' theory. This accepted that the state in France was able to intervene in the French economy to ensure its stability. By accepting the 'Common Programme', the PCF implicitly accepted the possibility of using the state towards socialist ends. Balibar, in criticising the PCFs new approach did not doubt that the state could intervene to restructure capital. The major question was whether it could intervene to genuinely promote social justice. In Balibar's view an affirmative answer suggested that the state was not an integral part of capitalism. The implications of the PCF's strategy were, in other words, social democratic.

5. As Parkin (1979, p191) has pointed out:

    the frequency with which the men on horseback have been called in to shore up crumbling bourgeois regimes plainly indicates that whenever a propertied class has faced a choice between loss of liberty and loss of property it has not usually been paralysed by indecision... Although modern social democracy does not aspire to the complete expropriation of private property, its strategy of gradually gnawing away at the institution could also meet at some point with unexpected resistance.

6. In practice of course, the notion of 'defensive violence' is a difficult one to apply, as the Austrian socialists found in 1934. On the one hand ignoring the issue can lead to a lack of preparation in any conflict, not of one's choosing. On the other the preparation of contingency plans play into the hands of those who argue that social democracy is not committed to the democratic path (Lesser, 1976).

7. This is less the case with the 'Austrian' branch of neoclassical economics which has emphasised the way equilibrium is continually disturbed. The importance of the market mechanism is seen to lie in its role as a carrier of information and in demonstrating the opportunities for profit and innovation.

8. This response can be seen in Jay's (1937) Socialist Case, Meade's (1948) Planning and Price Mechanism, and Lewis' (1949) Principles of Economics Planning. Similar developments were taking place elsewhere, notably in Sweden.

9. The later reappraisal of Keynes, associated with Clower (1965) and Leijonhufvud (1968), was an attempt to state that Keynes had in fact provided the general theory and that it was the neoclassical case of perfect information, where all agents trade at market clearing prices, that was the special case. This tradition stresses the importance of imperfect information, quantity adjustment to prices when trading occurs at nonmarket-clearing prices and the possibility of multiple underemployment equilibria. However for our purposes such an approach does not significantly differ from the Keynesian-neoclassical synthesis on prescribed economic policy. It does not provide a rationale different in kind to that promoted by traditional Keynesianism.

10. The roots of this position lie once more in Bernstein, and in particular his rejection of the 'breakdown' aspects of marxism. As Luxemburg pointed out, Bernstein did not only reject a particular version of breakdown but the very possibility of collapse:

for Marx's basic conception according to which the advent of socialism had its preconditions and objective roots within the process of capitalist production itself, Bernstein substituted a socialism based on the ethical ideal, the goal of civilised humanity free to choose its own future in conformity with the highest principles of morality and justice. (Colletti, 1972b, p54).

11. A distinction should therefore be drawn between political economy and left social democratic theory. While left social democrats adopt a political economy stance, it is clearly not the case that all those working in the political economy tradition support left social democratic policy.

12. For a range of articles within the political economy tradition that have developed this theme, see Goldthorpe (1984) and Lindberg and Mayer (1985).

13. Such theorists as Lindblom (1977) and Offe (see Offe and Range, 1976) have gone beyond functionalist Marxist theories of the state to explain the state's institutional and structural dependence on capitalist investment and 'business confidence'.

14. A response from an economics textbook in the neoclassical tradition will show what is at stake, as well as demonstrating that the differences entail ideological prejudices:

> the post-Keynesians' rejection of general equilibrium analysis in favour of their own methodology of treating the economy as a historical process explains why their approach has had much less impact on policy formulation than the neoclassical-Keynesian synthesis. Policy recommendations from this school are of the <u>broad brush stroke</u> variety and would require major elements of government direction in the economy and the replacement of market price signals by <u>bureaucratic</u> decision-making. (Rebmann and Levacic, 1984, p303, my emphasis).

15. As Eichner (1978) points out this 'incipient paradigm is more a programme of research' and does not entail a single conception of economic policy. A 'realist' analysis of capitalist economies is as important for the reformer as it is for those seeking more radical change.

16. As Allsopp and Helm (1985) have argued in a critique of orthodox economics:

    > there is a tension within economics between *a priori* theorising based on axioms of rationality and the logic of choice, and its claim to be an empirical science... it is hard to resist the conclusion that economists need to become more philosophically sophisticated and pay more attention to what used to be called political economy.

17. In Britain, Cole challenged the Fabian consensus of Keynesianism and the mixed economy as a permanent resting place: 'he thought that Keynesian economics was too involved with aggregates and not sufficiently concerned with the structural problems necessary for a socialist economy to replace the capitalist system' (Durbin, 1984, p49).

18. Left social democrats would point to the limited progress achieved in redistribution of income and wealth during the years of social democratic consensus. In Britain the work of Titmus was an early example of this dissatisfaction. Later studies by Townsend on poverty and LeGrand on the welfare state indicated the levels of poverty still remaining and the extent to which post-war policies had led to redistribution within classes rather than between classes (Plant, 1984; Wright, 1984).

19. The emphasis on circulation and the neglect of production is inherent in the structure of orthodox economics, both in pure neoclassical theory and Keynesianism. Marxist theorists have argued that a similar neglect is also in evidence in post-Keynesian, or neo-Ricardian, theories (Rowthorn, 1980, p41).

20. Murray (1984) has argued that the decline of the social democratic consensus has been related to important changes in capitalist production. Keynesianism was associated with the 'Fordist' model of production - mass production of standardized goods, use of specially designed machinery, production lines and a semi-skilled workforce. This was associated with a mode of consumption entailing higher wages to allow purchase of standardized consumer goods. Murray's argument is that the end of the economic boom can be linked to the crisis of Fordism. The monetarist response is seen in terms of changing the balance of economic power against workers but also in promoting modes of post-Fordist production. Thus for Murray the economic crisis is one of restructuring in which socialist economic strategies should intervene to change the course of that restructuring.

21. It is costly and difficult for unemployed workers to identify potential employers and vice versa; simple wage reductions will not automatically stimulate the demand for labour because most employers, quite rationally, are involved in long-term relations with their employees and it is again costly to disturb such relations. Whereas flexibility is built into full information markets, decentralised exchange requires a degree of inertia and internal rigidity to make it possible. (Grahl and Teague, 1990, p47).

The importance of seeing the labour market as a social institution has also recently been stressed by Solow (1990).

22. For instance, Hall in analysing the decline of the British economy in the twentieth century is critical of models that assume the efficiency of the market and look to exogenous variables, such as poor macroeconomic policy or state intervention, as the cause of the decline. While conceding that microeconomics has widened its scope of inquiry to incorporate such factors as barriers to entry and information costs, Hall (1986, p35) argues the need to recognize that 'markets are themselves institutions. In short, the market setting in which entrepreneurs and

workers operate is a complex of inter-related institutions whose character is historically determined and whose configuration fundamentally affects the incentives the market actors face.' For the British case, Hall examines the role of the financial-industrial nexus, product markets and the structure of industry, firm organisation and the structure of the labour market and shop-floor power. He concludes that because 'of the structure of British markets, the 'invisible hand' failed to transform the efforts of individuals to maximise their own welfare into long-term benefits for all. Instead, market incentives led managers and workers to take decisions that brought slow rates of growth and a deepening circle of economic decline' (Hall, 1986, p47). Hall uses such a foundation to examine why other states such as France and Germany were able to use market incentives and institutional intervention to promote a more successful modernisation and restructuring of their economies.

23. We have indicated that monetarism too is a form of intervention. As Gramsci argued in his Prison Notebooks:

> The ideas of the Free Trade Movement are based on a theoretical error whose practical origin is not hard to identify; they are based on a distinction between political society and civil society, which is made into and presented as an organic one, whereas it is a methodological one. Thus it is asserted that economic activity belongs to civil society, and that the state must not intervene to regulate it. But since in actual reality civil society and state are one and the same, it must be made clear that laissez-faire too is a form of 'state regulation' introduced and maintained by legislative and coercive means. It is a deliberate policy, conscious of its own ends, and not the spontaneous automatic expression of economic facts. (quoted in Hodgson, 1984, p78).

2 Indicative planning

The prestige of indicative planning reached its peak during the 1960s. Its popularity in Britain at that time reflected the widespread belief that it had contributed to the successes of the French economy. Furthermore indicative planning was linked to an elegant economic theory. As we shall see both these elements are in fact problematic, and by the 1970s the prestige of indicative planning had considerably declined. In Britain the very partial and spasmodic experiment was abandoned by both Conservative and Labour governments, and in France, where its practice had been more persistent and systematic, the role of the plan was increasingly peripheral to economic policy.

Here we examine the rationale, scope and limitations of indicative planning. We argue that indicative planning, at both the level of politics and economic theory, can be conceptualised in terms of the social democratic consensus, discussed in chapter one. At the level of politics it constituted the negation of more interventionist forms of planning. Masse (1965), the most influential French planner after Monnet, made it clear that planners sought to challenge neither the market system nor the profit motive. On the contrary the issue was to make these two elements of the capitalist system more effective. As Cohen

(1977, p4) has stressed, the achievement of Monnet in the immediate post-war era was to 'educate' French businessmen that planning, and with it the modernisation and rationalisation of industry, could serve their interests, and, indeed, given the political context, that it was the only alternative to more full-scale nationalisation and interventionist planning.

At the theoretical level indicative planning implied that the basic problem of the market economy was one of information: '...from a planning point of view, the model apparently implies that all the economy's information and incentive problems can be resolved once the right prices, namely the equilibrium prices, are somehow determined' (Hare, 1985, p116). As we argued in chapter one, the concern with equilibrium and 'right' prices is the hallmark of orthodox economics. Thus while theory provided a rationale for intervention, its form was very much within the boundaries of the social democratic consensus. Indicative planning was a useful accompaniment to macroeconomic policy, implying once more the need for intervention through the market and the abandonment of more interventionist methods.

It was the perceived failure of indicative planning and macroeconomic policy in the 1960s and 1970s to confront increasingly severe economic problems that led many socialist parties in the 1970s to propose more radical policies. By examining the scope and limitations of indicative planning we can provide a context for such alternative economic strategies which are discussed in chapter three.

In section 1 we discuss the economic theory of indicative planning, while in section 2 we look more closely at its practice in France. In section 3 we examine the political economy of indicative planning and some aspects of a left social democratic critique.

2.1 The economic theory of indicative planning

The economic theory of indicative planning is predicated on the existence of informational failure in the market system and an examination of the extent to which this market failure can be redressed by the dissemination of information to ensure a more efficient allocation of resources. As Estrin and Holmes (1983) point out such a concern has a theoretical antecedent in

Keynes:

> Many of the greatest economic evils of our time are the fruits of risk, uncertainty and ignorance... I believe the cure for these things is partly to be sought in the deliberate control of the currency and of credit by a central institution, and partly in the collection and dissemination on a great scale of data relating to the business situation, including the full publicity, by law if necessary, of all business facts, which it is useful to know. These measures would involve society in exercising direct intelligence through some appropriate organ of action over many of the intricacies of private business, yet it would leave private initiative and enterprise unhindered. (Keynes quoted in Estrin and Holmes, 1983, pp7-8).

Since individual agents in the economy hold different views about the future, in ex post terms this can lead to an inefficient allocation of resources if expectations fail to coincide. This becomes a more serious problem in a world of rapid technological development and with the long-run nature of corporate investment planning. Investors face incomplete information on which to base their investment decisions. A second problem concerns the uncertainty over the future development path of the economy which will also influence investment decisions. To confront these problems the state can promote 'generalised market research' to provide estimates of future prices, in order to make future expectations consistent, and reduce uncertainty by outlining feasible future development paths and/or the conditions under which any one path would be followed (see Hare, 1985, p118, p122). This can lead to an increase in investment and to the efficiency of decisions ex post. In the early literature the emphasis was on reducing possible bottlenecks: 'overall growth rate will not be held back by one (or several) key industries, which, because of over-cautious estimates of future demand, did not expand rapidly enough to meet the needs of the growing economy' (Masse, quoted in Cohen, 1977, p77).

For its proponents the beauty of this theory is its self-implementing aspects. Market agents will see where their interests lie and carry out the plan, especially since they themselves provide the information on which the plan is constructed. As Cohen (1977, p10) concludes: 'the motor of indicative planning is a benign circle: the more industry follows the plan, the more accurate the plan's information will be; the more accurate the plan's information, the more reason industry will have to follow the plan'.

While here the stress is on microeconomic consistency, another approach in the literature stresses the need to plan ex ante for the rate of growth. A government committed to a particular growth target will raise the expectations of businessmen of future demand and may persuade them that a faster rate is feasible (Beckerman, 1972; Collis and Turner, 1977). Here the link with Keynesian thought is clear in that expectations determine investment which in turn determines the rate of growth. Here too quantitative forecasts play a major role, and by being consistent will affect how individual decision-makers act.

Before examining the theory in more detail we can point to some initial problems for such a conception. Nuti (1986, p87) has delineated three crucial obstacles to the automatic operation of the model through individual self interest: participants may cheat, their views may not be open to summary by 'single-valued and firm expectations', and even if this is possible there may be no agreement on participants' individual part in the overall plan.

The 'rules of the game' in the model require agents to act as they stated they would - one can easily set up examples, in a game-theoretic context, where it would be in the interests of one party to break the rule if everybody else does not. Furthermore Stiglitz's (1984; see also Grossman and Stiglitz, 1986) work on the theory of information has shown that monopolies may have an incentive to give false information on the best neoclassical profit maximising grounds. Finally in the model as it stands, agents are free to use the information in any way they choose.

Lutz (1969) has argued that there is no simple way to aggregate individual views into a common view of the future. Echoing the concerns of the Austrian school of economics, she argues that the market system works by rewarding those who make the best estimates about the future. Moreover if planners have misjudged the future, we risk an actual deterioration in resource allocation and an inefficient investment portfolio.

Richardson (1971) considers the extent to which branch forecasts of any plan will be useful for individual firms. Lutz (1969) had feared that this could be solved in practice by an unhealthy collusion by the major firms (see section 2.2).

Richardson argues that in a perfectly competitive environment, it is impossible for price information to provide a firm with knowledge of its share - hence the development by Edgeworth and Walras of provisional contracts and iteration before final prices were given by the auctioneer. The central problem is that there cannot be a complete set of future markets because producers cannot predict their future possibility sets and consumers cannot know their future demand. While Richardson (1971) accepts that planners could start with information from the basic unit, he argues that the more disaggregated the plan, the less its credibility will be in that it cannot hope to 'weld together the expectations of each and every entrepreneur into a consistent plan'.

Meade (1970) has attempted to take on some of these criticisms with a sophisticated and elegant response. Crucial to his analysis is a distinction between environmental and market uncertainty. In his example of a firm producing umbrellas and sunshades, the firm is uncertain on both the state of the weather, but also the likely demand given the state of the weather. He attempts to show that many of the problems above are concerned only with environmental uncertainty.[1]

Meade's (1970) model mimics the general equilibrium model of Arrow-Debreu with a full set of future contingent markets. He begins with a model of perfect competition, no externalities or public goods, no economies of scale, equal income distribution and so on. Instead of a Walrasian auctioneer, every agent in the economy is invited to the Albert Hall to start the tatonnement process. The auctioneer gives current prices and asks the assembled to write down for every day in advance supply and demand of goods and services at these prices (including of course how much they would borrow/lend at any future rate of interest). The process of iteration should come up (given a further set of assumptions, for instance, the perfect substitutability of all goods) with a vector of prices (including future prices) to clear all markets. The indicative plan becomes a perfect substitute for a full set of forward conditional markets. This implies that there are as many plans as there are possible paths of the economy. Here the inadvisability of a 'common view of the future' is countered for:

> it would leave complete competitive freedom for economic agents to act on their own diverse assessment of future changes in the environment... but at the same time it

would remove unnecessary uncertainties about the combined market implications of the diverse plans of these free economic agents. One would know, given the diverse views of plans of the various economic agents, what these implied for the future development of market conditions along each environmental path. (Meade, 1970, p29-30).

Meade goes on to discuss the implications of relaxing some of the assumptions of the model and we cannot hope to do justice to the whole breadth of his discussion. For our purposes it is enough to stress that transactions costs prevent anything like the calculation for market uncertainty under every environmental path.[2] This restriction entails the existence of a certain amount of 'residual uncertainty'. Meade contends however that the project is of some use since market agents could use representative time paths as a bench-mark for making their own time paths. Of crucial importance will be how close the true environment is to the representative one, and the extent to which entrepreneurs are influenced. Firms may well disagree about the planner's environmental path(s) and use the plan information quite selectively, combining it with other available information. Here the self-implementing feature of the plan is threatened (Hare, 1985, p92). As Estrin and Holmes (1983, p18) conclude: '...in general, for a given level of transactions costs, the more the sources of unpredictability in the economy, the less useful Meade's exercise will be. We risk what Lutz feared - a convergence of views on the wrong outcome rather than the right one, hence worsening resource allocation'.

Thus Estrin and Holmes (1983) point out that French planning was seen to be most effective in the 1950s and 1960s when the economic path of the economy was relatively stable and predictable. But by the late 1960s the increasing uncertainty of the world economy entailed an increase in the number of possible environmental paths and planners made serious errors in their predictions. Although this could have led to resource allocation problems, in actual fact they conclude that the net result was that fewer people took notice of the plan: 'the institution designed to 'reduce uncertainty' lost its importance in a period of rising uncertainty' (Estrin and Holmes, 1983, pp18-19).

Estrin and Holmes (1983) hold a middle position between the indicative planners and those of the Lutz position. But they do not believe that Meade has provided the correct rationale for indicative planning. Their case is based on the existence of economies of scale in the information field and the concept of bounded rationality. They point out that in the Meade model there is no reason for forward markets not to exist - there exist opportunities for profits which have not been exploited since information would be privately beneficial. The problem is whether the factors preventing the existence of future conditional markets will also prevent the functioning of an indicative plan. The answer they suggest is yes if planners try to do too much.

In particular, they point to the limited way in which uncertainty is treated in the Arrow-Debreu model. The treatment of information as fixed and immutable is unrealistic for: '...if people use information about each others behaviour as well as what they discover for themselves about the environment, market activities automatically generate externalities because each person's behaviour affects the information that he and others possess' (Estrin and Holmes, 1983, p28). In this case there is no guarantee that optimal decisions will be reached, either because the information is not available or people are unaware of its significance. Information has a public goods property - additional information which could benefit everybody but depends on costly actions of one individual may well not be transmitted (Grossman and Stiglitz, 1976). As Hare (1985, p121-2) has argued in the context of firms who need information about each others behaviour, or where firms need to identify possible trading partners, there is a further rationale for some form of indicative planning to reduce transaction costs.

'Bounded rationality' implies that individuals and firms simply do not have the capacity to examine more than a certain number of options and this: '...obstructs the definition, let alone the attainment, of private or social optimality in the allocation of resources' (Estrin and Holmes, 1983, p29). In other words, firms simply do not have the capacity to generate, or assimilate, the information that would be needed in a Meade-type plan. Planners therefore have a role in providing information which may have been overlooked, for instance because of their knowledge of world market trends, or because of the importance of public sector activities in certain investment decisions (see also Hare, 1985, p123). Thus for Estrin and Holmes (1983) the existence of 'bounded rationality' and economies of scale in the gathering and processing of information provide a rationale for a very modest form of indicative planning. Rather than attempting to impose a unique view of the future such planning should seek to broaden the opportunity set of decision-makers by disseminating economic information (Estrin and Holmes, 1983, p29, p51). Furthermore as market agents would be free to make what they want of any such information: 'the procedure should provide enough information for agents to avoid the most serious inconsistencies if they wish, but would contain no controls or imperatives' (Estrin and Holmes, 1983, p52). Their thesis then is that in trying to do too much the French planners eventually could not cope. This led to mistakes which discredited indicative planning and with it the limited, but useful, role it could have played.

For both Meade (1970) and the more limited form of indicative planning envisaged by Estrin and Holmes (1983), it is the market, and the individual decisions of market actors, which are the final arbiters of economic decision making. Planners can merely provide a level of information to promote the more efficient allocation of resources, with the most significant point in contention being the extent to which planners have the capacity to improve on the market allocation of resources.

2.2 The French model

We have seen at the theoretical level some of the obstacles that confront a conception of indicative planning which consists of the promotion of a consistent and consensual framework for future developments carried out by the operation of the self-interest of the participants. We can develop the argument by examining the French experience with planning in the post-war era. While this experience with planning constitutes the most persistent and systematic exercise in indicative planning, the French 'model' was never a pure one. As we shall see, the plan encompassed a diverse range of activities which were never fully coordinated into a coherent whole - the plan worked in several ways and at several levels. Delors (1978, p10-11) shows that French planning in practice contained five goals: the study of the generalised market, the reduction of uncertainty, the improvement of the allocation of resources, the creation of a framework for social debate and the education of the French on the need for modernisation. It is clear from the previous analysis that indicative planning cannot achieve all this - indeed a 'pure' indicative plan cannot orientate the development of society.

Before examining the range of activities covered by the plan we can provide a brief and schematic outline of the French planning process. For convenience this is usually discussed in terms of four distinct phases: analysis, dialogue, formulation and implementation (Hare, 1985, p89). Responsibility for the plan rests with the planning commission (Commissariat General du Plan - hereafter CGP) which was created outside the framework of the traditional government ministries. We cannot therefore equate the plan with the simple expression of state economic policy. During the operation of one plan, the CGP carries out a study of its operation in preparation for the next plan. It reports to the government on the results of these studies in terms of future short and long-term economic problems, possible bottlenecks and so on. The CGP then asks for an orientation from the government - the drawing up of options including a choice of, say, a 5.5%, 6.0% or 6.5% rate of growth. The crucial decisions are taken here in tripartite negotiations between the CGP, the Ministry of Finance and the government (Cohen, 1977, p35). The CGP is a small organisation with funds for administration and research, while the purse-strings are firmly in the hands of the Ministry of Finance which controls the budget and short-term economic policy. If the planners are to be effective, the backing of the Ministry of Finance is crucial.

The outline plan then goes to the Modernisation Commissions, where the <u>economie concertee</u>, stressed so much by Masse (1965), takes place. In order to promote social debate these commissions consist of representatives from business, unions and pressure groups. All are appointed by the government, on the advice of the CGP, but with no fixed proportions of seats. For in line with the emphasis of indicative planning on information exchange and the self-implementing aspects of planning: 'no rigorous dosage is demanded in the composition of the commissions. The spirit counts more than the letter. The commissions seek a general accord and not a majority vote. The goal of their work is not to separate the winners from the losers, but to come up with a common view' (Masse quoted in Cohen, 1977, p60). The basic idea is that the plan undergoes an iterative process within the commissions to get a coherent set of targets to be presented in an input-output table form, with vertical commissions working in particular sectors and horizontal commissions striving for coherence. The deliberations at this stage could lead to new data or inconsistencies which would entail a further iterative process to reach a consistent plan, and thus the second and third stages of the plan tended to merge (Hare, 1985, p90).

By the time the plan is approved only a single variant remains, usually in the form of a general equilibrium resource allocation model of an input-output table. But the existence of such a table can be misleading and, as we shall see, one cannot assume that this constitutes the heart of the plan. Nor can we assume that planners expected the plan implementation stage to be as automatic as theory would suggest. To see this we must go beyond the above rather 'ideal' schema of the planning process.

French five-year plans did not all share the same objectives, reflecting both the changing nature of the economic problems confronted and the perception of planners concerning the appropriate scope of planning to address such problems. The initial approach in the 1940s and 1950s focused on economic growth and reconstruction, on removing supply-side bottlenecks and modernising six major sectors of industry (Hall, 1986, p146). By the 1960s the scope of the plan was being broadened with the emphasis switching to enhancing French competitiveness in a more open economy and promoting harmonious development. Thus the fourth plan (1961-66) was more ambitious, approaching an overall resource allocation model. The whole economy came under the purview of the plan, including targets of growth for public investments and an attempt to define a national consensus of social classes and groups with a commitment to increasing spending on social infrastructure. In the mid-1960s French planners introduced value planning to replace projections in volume terms, implying that from the plan one could now read off the incomes and prices needed for the realisation of economic and social objectives. This led, as we shall see, to incorporation of incomes planning into the process. Thus 'planning in France has gradually shifted away from the production of co-ordinated branch plans covering the whole economy towards macroeconomic planning, largely in financial terms, combined with special programmes to meet the particular needs of the coming plan' (Hare, 1985, p91). This had a contradictory consequence. On the one hand resource allocation planning was more ambitious. On the other hand this entailed more accurate forecasts on the future developments in all sectors of the economy. As the economic conditions of the late 1960s and 1970s deteriorated this became an increasingly 'fragile' exercise (Hall, 1986, pp146-7).

To examine what the effect of planning in France was, we must look further into what it entailed in practice. While the <u>economie concertee</u> conception seems to fit in nicely with the theory of indicative planning, the actual practice was rather different. As Cohen (1977, chapter 6) has argued it is better understood as a tripartite bargaining process between the CGP, the government and big business. The planners' traditional concern with rationalisation and modernisation is aimed at industry, which in turn controls most of the investment in the economy. It is unrealistic to assume an open process of information exchange, let alone a pluralist negotiation on all priorities, when economic power is so unevenly distributed. Big business entered the negotiations with powerful weapons on its side: the ability simply to ignore the plan, independent access to other centres of power within the government and a quasi-monopoly over information. Trade unions had a very peripheral role within the <u>economie concertee</u> (Hare, 1985, p90).

The power of industrialists within the <u>economie concertee</u> leads to a practical solution to the problem, highlighted by Lutz (1969) and Richardson (1970), of market shares. Most analysts point out that CGP negotiations take the form of one member of the CGP talking to a particular industrialist in the monopolistic (or oligopolistic) sectors or a representative of a trade association. Many vital decisions are not arrived at in the plenary sessions of the commissions. Furthermore as Cohen (1977, p72) concludes: 'it is absurd to assume that a group of leading businessmen in an industry who are assembled together to prepare a set of investment, output and modernisation programmes for industry will not continue to act co-operatively when the time comes to break down that plan into programmes for each firm'. One hardly need add that such collusion does not necessarily improve the allocation of resources.

Neither have planners taken the stance of Olympian detachment suggested by the theory of indicative planning. In the early post-war period when investment finance was scarce, planners through their channelling of Marshall aid and control of import licenses had considerable leverage over the planning process (Hare, 1985, p93). The government also intervened more readily. For instance in 1951 after failing to convince steel producers to respond to the planners' wishes direct intervention to change the steel cartel was carried out. But as Delors (1978, p15) concludes: 'they were the early days in the French planning experience. Over time, the use of the stick became increasingly rare. Today it has virtually become a museum piece'.

However an element of dirigisme has always been an integral part of French planning. In the 1960s when the concern of planners shifted to the international competitiveness of individual firms, rather than planning for whole sectors, the form of this dirigisme altered. The shift was to introduce selective measures aimed at individual firms (Hall, 1986, p167). Hall mentions a number of instruments available for promoting industrial concentration and encouraging French leading firms to compete on the international stage: bilateral contracts, tax advantages, export assistance, price control exemptions, investment funds and bank lending policies. Nor was this process any more open or transparent than the conception of economie concertee: 'to implement the new industrial policies, however, planners often had to negotiate directly with the managers of the firms they were intending to support, without the mediation of trade associations, who frequently opposed rationalisation in order to protect their weaker members... Such agreements could only be reached behind closed doors' (Hall, 1986, pp168-9).

If this relationship between government and industry was at the core of French planning there were other factors which mitigated against the plan acting as an overall resource allocation plan, that is as the focus of all economic activity. We can briefly mention three such problems: the inability to incorporate short-term economic strategy, the failure to plan the public sector and the increasing intervention in the private sector outside the structure of the plan.

Given the analysis of section 2.1, it is not surprising that most French plans were interrupted by balance of payments or other economic problems leading to short-term stabilisation plans. The second plan (1953-1957) attempted to incorporate the short term by introducing the balance of payments as a major target. However this was interrupted by the short-term deflationary package, known as the Rueff programme. The fourth plan was also interrupted by a crisis, leading to a similar package introduced by Giscard, then Minister of Finance. Both of these stabilisation packages did not entail the abandonment of their respective plans. But by the late 1960s, when as we have indicated there was a greater reliance on financial planning and thus on accurate economic forecasting, the lack of coordination between short-term economic measures and planning was more serious. By 1973 with the first oil crisis, macroeconomic policy was not even disguised as being an integral part of the plan. The seventh plan (1978-81) aimed at full employment and the reduction of inflation and the balance of payments deficit. However there was no serious attempt to operate a consistent macroeconomic strategy. Increasingly plan macroeconomic forecasts began to be a source of embarrassment, whereas in the preceding period the relative success of growth and other indicators added considerable prestige to government economic policy makers. Eventually Barre was to claim that such forecasts should be treated as just one contribution to the pluralistic availability of information (Estrin and Holmes, 1983, p100).

The failure to incorporate macroeconomic strategy reflects a number of factors which cannot be expanded fully here. The proximate cause was the traditional conflict between the Ministry of Finance and the planners, with the former having a more powerful and central role in economic decision making. But such conflict also reflects different conceptions on the appropriate response to economic problems and the relative importance of short as opposed to longer-term economic measures. As Cohen (1977, p262) argues, there exists:

> a fundamental difference between the political forces that converge upon the existence of short-term policy and those that had stakes in the plan... Just because the fifth and sixth plans adopted a comprehensive planning approach that depended on a tight coordination of short-term policy with the middle-term plan is no reason to assume that the political forces that had made such coordination impossible in the past would tactfully vanish. They didn't.[3]

Thus by the 1970s the short-term economic policy of the Ministry of Finance, and the fight against inflation had become the central economic concern of the government, thereby considerably relegating the scope and prestige of planning (Estrin and Holmes, 1983, p151).

Here an important distinction should be made. The new economic 'liberalism' of Giscard and Barre in the 1970s did not entail the abandonment of dirigisme. The French government continued to intervene in both the private and public sectors in the 1970s. But increasingly this took on an ad hoc nature and it was outside the framework of the plan (Delors, 1978, p27). By the 1970s the CGP, as well as the Ministry of Finance, faced a new challenge to their power in the form of the Ministry of Industry. It was the Ministry of Industry which increasingly was responsible for this new intervention and which was given powerful instruments such as block grants to influence events.[4]

Finally the coherence of planning was undermined by the fact that the plan never fully incorporated the activities of the public sector. The fourth plan tried to incorporate the public sector and there was a new emphasis on social expenditure. Once more the only method open to the CGP to ensure that social spending on programmes such as transport, urban planning, education and health was carried out, was its lobbying activities with the Ministry of Finance. In fact it was relatively successful with plan fulfilment targets in particular areas ranging from 85% to 95% (Delors, 1978, p20).[5] However such attempts were spasmodic and incomplete and were quickly being overcome by events. Increasingly planners were seen as merely one pressure group, in favour of social expenditure. By the time of the fifth plan the planners were seen to be presuming too much to dictate social policies - the government actively intervened to deprioritise social expenditure.

Planners had similar difficulties in influencing the nationalised industries. In the 1950s it often seemed easier to influence private firms. This led to the publication of the Nora report in 1967 which recommended that <u>contrats de programmes</u> should be signed between the state and nationalised industries. In fact only two were signed. Thus while in 1973 the government had a potential to influence capital formation directly (over 15% of gross fixed capital formation was carried out by non-financial public enterprises) it never seems to have used it share of productive investment as part of the overall plan (Estrin and Holmes, 1983, chapter 6).

From the above it is clear that by the mid-1970s planning in

France had reached a serious decline. Yet in the 1950s and 1960s most commentators consider that it was a relatively successful exercise. It did help rationalise and modernise French industry, and provide information to economic actors which would not otherwise have been available. The extent to which it contributed to the successes of the French economy before the 1970s is more problematic.[6] For we have seen that both the theory of indicative planning and its practice in France suggest that the plan was never a self-implementing process. In the French planning process it is hard to distinguish the effect of purely indicative aspects from the associated, and more dirigiste, forms of planning. Whether French planning can be considered a satisfactory model by left social democracy is discussed in the following section.

2.3 The political economy of indicative planning

At the level of theory indicative planning is based on the premise that the market allocation of resources, apart from some imperfections such as uncertainty and informational failure, is desirable and leads to economic efficiency. As we argued in chapter one this necessarily leads to a particular conception of the nature of economic problems and thus to the scope of economic policy. Thus indicative planning constitutes a natural accompaniment to Keynesian demand management and intervention in the economy when, and if, market failures exist.

While French planning also incorporated elements of dirigisme into the planning process, it too accepted the market as the dominant economic nexus. While one of the goals of French planning was to broaden social debate on economic policy, in actual fact it accepted as a starting point the unequal structure of power and advantage within the market. While the economie concertee essentially came down to a bargaining process between the state and industrial sectors, the shift to bilateral agreements with individual firms to promote dynamic leaders entailed an even stronger relationship between state and certain sections of capital. As Hall (1986, p171) has pointed out: 'under the aegis of the planning process, some segments of industry acquired a degree of influence over the nation's industrial policy that bordered on control'. As we argued in chapter 1, such a reliance on the private sector, or sections of it, necessarily limits what any planning exercise can achieve.

Indicative planning cannot in practice be about merely 'efficiency' or 'technical' problems. It too has important political economy implications. To shed further light on this, and thus set the scene for left social democracy's more interventionist stance, we focus on two questions. Firstly, we examine the extent to which indicative planning can confront institutional or structural obstacles to economic development. And secondly we discuss the extent to which indicative planning promotes genuine social debate on economic priorities and future economic development.

We have seen the difficulties involved in assessing the extent to which French planning contributed to the successful economic development of the French economy in the 1950s and 1960s. French indicative planning was working within a context of high economic growth. Lamfalussy has made an important distinction between the different investment strategies of firms in low and high growth contexts (see Holland, 1972, pp14-15). To summarise very briefly, in a low-growth context firms are likely to suffer a 'penalty' effect if they follow an indicative planning target within an industry when conservative firms do not. However in a high growth context the 'penalty' is more likely to be felt by firms refusing to follow a particular target, while others do, and thus risk losing market shares. Holland (1972; 1975) has argued that the success of French planners was based on contingent factors and that countries grappling with the problems of low growth have to rely on more interventionist policies. A central question is whether indicative planning is of much use during a recession, if the private sector will not invest, or in the context of an economy with structural problems facing long-run economic stagnation or decline.

The brief British experience with indicative planning in the 1960s highlights this issue. The popularity of indicative planning at that time reflected the concern over Britain's relatively poor economic performance and the effect of 'stop-go' demand management on investment fluctuations (see Hare, 1985, pp31-3). Significantly it was felt that France's relatively superior economic performance reflected the self-implementing aspects of indicative planning, ignoring the effects of the more dirigiste aspects of French planning. In 1962 a new planning body was set up. The National Economic Development Council (NEDC) consisted of representatives from government, unions, employers' organisations and nationalised industries and the Director-General of the National Economic Development Office (NEDO). Its first plan (1961-1966) was based on a 4% growth target (much higher than anything reached since 1945) and was based on 17 industrial branch studies where the industries had been asked to specify obstacles to achieving the growth target (Hare, 1985, p34). In 1964 the Labour Party reaffirmed this new commitment to planning by setting up the Department of Economic Affairs (DEA) to coordinate planning and act as a counterweight to the Treasury. While the DEA was now responsible for the new National Plan, it cooperated with the NEDC to provide a British version of the <u>economie concertee</u>. The National Plan included once more a high growth target of 3.8% and attempted to break down the implications of this for various industrial branches. However the process was not as rigorous as in France and it provided very few indications on the possible obstacles or bottlenecks to its implementation and suggested that by improving the availability of information this would lead to a self-implementing process (Hare, 1985, pp38-9).

The widely acknowledged failure of the National Plan, and the eventual demise of the DEA reflects a number of factors. Firstly the plan was very vague on the institutional and structural features which may have led to Britain's poor economic performance. Indeed as many observers have pointed out given the lack of such an analysis, what was surprising was that Britain was not already performing better. Thus any obstacles, such as the balance of payments, were wished away merely by assuming that exports would grow at twice the rate they had done previously (Gilhespy <u>et al</u>, 1986, p90). As Hare (1985, p39) has pointed out in the context of a commitment to a high exchange rate for sterling this was clearly problematic. This suggests that a reliance on information exchange and a demand-expectations version of indicative planning suffers from an underestimation of the supply-side problems of the economy. In Britain market actors simply did not believe over-optimistic growth targets and thus altered neither their behaviour nor their expectations. There was no attempt to confront economic problems as a structural issue.

Lacking an analysis of the structural problems of the economy, the National Plan could only fail in providing a vision of the direction in which the economy should be heading: 'the exchange of information was elevated above any conception of what should be done. No mechanism was envisaged to change the behaviour of companies - the focus of decision-making power in the economy -and in particular to improve the level and quality of investment' (Eatwell and Green, 1984, p194). Nor did the National Plan incorporate any novel policy instruments: 'individual firms received little or no guidance as to what was expected of them, and the plan itself provided neither compulsion nor incentives (eg based on taxes and subsidies, as in France). Consequently the notion of a self-fulfilling prophecy made no sense in practice' (Hare, 1985, p39).

In chapter one we argued that central to the rationale of a left social democratic perspective, was an analysis of the structural and institutional obstacles to development. Thus, for instance, Gilhespy et al (1986, p17) argue that: 'the concern with overall demand and indicative planning has, by and large, left corporate strategy untouched. The contrast with Germany and Japan, for example, could not be more marked: government there has intervened directly in industrial reorganisation, and the links between industry and finance capital have been stronger'. The left social democratic alternative approach, which incorporates institutions and structural features, is the subject of chapter 3.1 where we discuss 'strategic planning' as an alternative to indicative planning.

The demise of Labour's National Plan was also a result of the failure to integrate it with short-term economic policy. In 1966 economic pressures led to deflation to control imports and in 1967 there was a devaluation. Apart from the other factors which we have pointed to, this failure to incorporate macroeconomic strategy in both France and Britain, reflects the failure to integrate planning with a policy on incomes. This leads us to the second question we are focusing on here, namely the ability of indicative planning to promote social debate and consensus on economic priorities.

In the French case the first attempts to incorporate incomes policies into planning occurred in the 1960s when French planners introduced value planning to replace projections in volume terms. The CGP held a conference on this policy in 1963, inviting all interested parties. As Delors (1978) notes the unions seized upon the change to translate the rhetoric of social consensus into some kind of reality. They proposed four conditions in exchange for an agreement on wages: taxation reform, a share for labour in the benefits of national economic expansion, tripartite negotiation on price and competition policy, and a transformation of the system of collective bargaining. As Delors (1978, p22) concludes:

> when the government grasped the implication they were taken aback. As a result they quickly enough consigned such terms to the cemetery of non-applied agreements... the negative result of rejecting the chance to negotiate with the unions on prices and incomes was crucial not only in itself but also - which has been neglected by many commentators - for the institution of planning. It was a key factor in the decline of planning in post-war France.

Thus French planning was never able to promote a consensus on economic priorities. By the late 1960s the left in France responded with the publication of a counter plan (see Cohen, 1977, pp214-19). And by the 1970s the Common Programme of the left proposed more radical solutions outside the boundaries of indicative planning.

In the British case there was also an attempt to incorporate incomes into planning with the creation of National Board for Prices and Incomes (NBPI) in 1965. This sought some consensus with unions on wages to be compatible with growth targets. Increasingly however as such consensus was lacking the government resorted to statutory methods, although this too needed some level of cooperation. By the late 1960s the attempt to promote a prices and incomes policy had to be abandoned (Hare, 1985, p38).

To many the failure of the National Plan was a result of the Labour government's refusal to devalue early enough to give some realistic chance of adhering to the growth targets. Rowthorn (1983) convincingly argues against this analysis. He notes the strength of British trade unions and the tendency of planning to legitimise socialist values. In this context planning, even in a mild form, may have had dangerous implications for capital. Thus Rowthorn (1983, p68) concludes:

> the abandonment of growth and full employment by the Labour government did not occur, as many believe, because the City triumphed over industry. It occurred because no significant section of capital was willing to support the kind of state intervention required to avoid deflation and maintain economic growth, and because the Labour government lacked both the will and the popular

support to implement such a programme against the opposition of capital and its allies at home and abroad.

In Britain too, the perceived failure of indicative planning and macroeconomic policy to sustain growth and employment led to a rejection by unions of attempts to incorporate incomes policies into government policy and also to the more interventionist policies that the Labour Party was debating in the 1970s.

The roots of the explanation of why indicative planning has failed to promote consensus over economic decision-making lie within its own theory. For behind this theory: 'lies the familiar assumption of contemporary economics that there is an essential harmony of interest among the actors in the socioeconomic field: conflicts among these actors arise primarily because they lack full information about their options and interests' (Hall, 1986, p159). A full resource allocation plan entails that the plan becomes the focus of economic activity for all agents in the economy.[7] If this further implies a debate at the level of society over future economic priorities, then this must entail a degree of conflict unless our model of society is one of perfect harmony. For a full resource allocation plan by linking all areas of the economy can be used to discover the implications of changes in particular areas throughout the economy. Such planning may well threaten the cosy nature of the economie concertee by increasing the range and effectiveness of collective decisions:

> as the scope of collective decisions expands, the composition and distribution of goods, services, efforts, rewards - even status and values - are likely to change. An explicitly planned income distribution is very likely to differ from present distribution. Profits might be threatened, so might business' sense of independence and power. The proportion of total output consumed collectively might increase. (Cohen, 1977, p162).

It is for this reason, Cohen (1977, p161) argues, that indicative planning theorists have emphasised the consistency of the plan as opposed to its explicitness:

> they fail to examine the potential for conflict between a planning process designed to increase the range and effectiveness of collective decisions and one designed to increase the efficiency of resource allocation by increasing

the coherence of decisions made by individual firms. Efforts to democratise the planning process reveal the existence of important trade-offs between the democratic control of a plan and its coherence.

Delors (1978) considers the problems of explicitness as crucial to understanding the demise of French planning. He argues that in the 1970s the governments of the Right needed a free hand. The social and economic base of the Right included both the large internationally orientated firms (whose interests were broadly reflected in the Barre/Giscard axis) and the more nationally orientated and smaller firms. Within this context:

> the government and conservative power had need of the first to assure both their policies and their economic prosperity. But they had absolute need of the latter in order simply to stay in power via the vote: ie to have more votes than the combined forces of the Left. This was the cruel dilemma, trying to reconcile the traditional middle class with the view of industrialisation as perceived by the technostructure and the new ruling class. The Plan was not a good means of doing this, since it means transparency and coherence in objectives. (Delors, 1978, p23).

The widening of the possibilities through a discussion of economic priorities must eventually lead to a subsequent narrowing, where there will be clear winners and losers. This is disguised by the theory of indicative planning but in practice, if the plan is even to pretend to promote some form of consensus at the level of society some sort of solution is needed. In the French case, Hall (1986, p160) argues:

> the real purpose of the Plan became to narrow the choices being actively considered - by excluding proposals held to be unfeasible in a market economy, incompatible with high levels of growth, or antagonistic to the intensive development of advanced sectors of industry - while appearing to widen them. This capacity to maintain an appearance of doing one thing while actually doing its opposite was what ultimately accounted for the Plan's great effectiveness as an agency for the mobilisation of bias.

For as we argued in chapter one it is in the very structure of orthodox economics, on which the theory of indicative planning is based, to limit the scope of economic policy.

The above political economy of indicative planning leads to some important conclusions. To some the problems of indicative planning, at both the level of theory and practice, suggests that the case in favour of planning should be reformulated on a less ambitious basis. Thus Estrin and Holmes (1983) argue that French planners failed in attempting to achieve too much. Instead planners should limit themselves to a more modest task of providing information and expanding the possibility sets of market actors.

For left social democracy the analysis of chapters one and two suggests a directly opposite conclusion. Planning must involve firstly an institutional/structural analysis of the obstacles to economic development and to the promotion of socialist goals. Secondly, planners need mechanisms and instruments to intervene in those institutions and influence the decision-making of firms. Finally, and perhaps of most importance, planning must entail an exercise in political economy. If planning is to entail democratisation, if democracy is to be extended to the economic sphere, then this process cannot be merely a technical question. The explicitness of such planning entails the existence of both winners and losers in the ensuing process. As we have suggested this further entails both coherent economic policies and a level of support within society for such a project. The coherence of the left social democratic alternative, its widening of the scope of economic policy and its more interventionist stance, is the subject of chapter three.

Notes

1. Masse was also sensitive to environmental uncertainty. He argued that there would have to be periodic adjustments of projections in the light of random events. One method he proposed for dealing with the problem was an inflation threshold after which there would need to be a procedure to find a new strategy appropriate to the circumstances. He adds that this would also have the added bonus of economic agents being warned in advance of a possible change in strategy, and with the proximity of any threshold, they may be induced to behave more moderately. Such a conception is a theoretical antecedent to the use of 'conditional precommitments' which are discussed in chapter 3.5.

2. Meade (1970, p37) notes that in his industry example if there are three values in each Year, and if we are dealing with a five-year plan, then there would be 14,352,807 different time paths for environmental uncertainties.

3. Such an approach can be usefully compared to our account of the political economy implications and rationale of monetarist economic policy discussed in chapter one.

4. As in Britain, private sector companies are eligible for a wide range of aids. There is no co-ordination of these: companies unable to get money from the Industry Ministry can turn to the Ministry of Finance. The state typically provides aid in the form of loan stock, has deliberately not taken voting rights. (Gilhespy et al, 1986, p94).

   An alternative and more interventionist strategy which seeks to coordinate state financial incentives and provide voting shares stock to operate 'commercial leverage' over private sector firms is discussed in chapter 3.2.3.

5. Delors (1978) is surely right to criticise Lutz (1969) who makes much of these figures to show a lack of success - the overall figures are less important than the shift in resources and expenditure.

6. In Hare's (1985, p93) opinion:

> its indirect effect on confidence and its role as a forum for interest groups to meet and exchange ideas were both extremely valuable benefits though most researchers have found it very hard to discern more concrete indications that plans for particular branches had much impact.

7. If the plan does not become such a focus then the government cannot commit itself and must respond to forces acting outside the plant (see Cohen, 1977, pp158-9).

3 Planning and macroeconomic control

In chapters one and two we have seen that, through its adoption of the political economy perspective, left social democracy has challenged both the means and ends of economic policy, thus entitling us to speak of alternative economic strategies. We have also examined the case, defended by left social democrats, for going beyond macroeconomic management and indicative planning. Here we examine if there also exists a coherent package of institutions and mechanisms, which could be said to constitute such an alternative economic strategy.

In section 3.1 we discuss the scope and formation of a national plan. While such a plan is common to both indicative and interventionist planning, certain distinctions in the overall approach will be drawn.

Section 3.2 goes on to examine what is meant by going beyond indicative planning to ensure the implementation of the plan objectives. This entails discussing the range of possible forms of ownership, as well as working out the appropriate balance between private and social ownership. Such institutions and mechanisms as state holding companies, new public enterprises, local enterprise boards and planning agreements, which have been promoted by left social democrats, are discussed at some length.

Section 3.3 is concerned with intervention in those institutions and areas which are considered important for the overall framework of development. Here we focus on the importance of finance and financial institutions; international trade flows and multi-national corporations; and competition policy.

Section 3.4 examines the relationship between planning and industrial democracy and the developmental state. This includes a discussion on whether workers' participation can be extended in a manner that is not dysfunctional to the other main areas of concern.

Section 3.5 attempts to bring together some of the themes in the above sections to examine the coherence, or internal consistency, of the overall package. We also point to any existing areas of inconsistency or ambiguity within the overall model.

Sections 3.6 and 3.7 examine how such a package fits in with macroeconomic policy, or short-term economic policy, including an extensive discussion on the planning of incomes.

Our goal in this chapter is therefore to work towards extracting from the literature a planning schema incorporating a range of institutions and mechanisms which have typically been supported by those in the left social democratic tradition. Such a schema is presented in 3.8 and can usefully be compared to that presented in Gilhespy et al (1986, p115) and Hare (1985, p260).

3.1 The national plan and the scope of planning

A national plan is common to all models which seek to introduce elements of planning into a market economy. Its function is to facilitate debate within society about alternative social, political and economic options; to delineate certain central economic and social priorities for medium to long-term development; to provide a framework for the co-ordination of economic decision making within society; and to clarify the framework of state intervention.

A central question concerns the scope of the plan. We do not examine in any detail the theoretical discussion concerning the appropriate balance between 'plan' and 'market'. Indeed in chapter one we suggested that the formulation between the 'plan' and the 'market' at an abstract theoretical level was not a particularly useful one for economic policy. Suffice it to say that there is now an overwhelming consensus that the plan is not intended to replace the market mechanism or directly determine the countless number of economic decisions made within the economy. Thus markets will continue to have a central role in the structure of output and relative prices in the various alternative economic strategy models being discussed here. As one of the more recent contributions has made clear, markets 'offer a framework in which planning for the future can start from 'actually existing' demand and cost parameters, and not merely from abstract, conjectural notions about what is 'needed' by society and about what is an 'adequate' level of efficiency' (Desai et al, 1988, p78).

One of the central tasks of most national plans is to attempt some form of forecast of future economic developments. Here too there is a need for some flexibility. For instance, Hare (1985) argues that at the macro level it is not particularly useful for planners to map out point forecasts. Indeed in chapter two we discussed the difficulty of planners in France predicting an increasingly unstable future. Hare prefers the scenario approach which can, for instance for particular projects, point to the existence of a critical range of, say, exchange rates.

It is usually assumed that planning must directly influence such macroeconomic variables as employment, aggregate income, aggregate investment and its broad allocation, public consumption and the balance of internal payment flows for capital and trade (Nuti, 1988, p84). Thus planning is geared to both macroeconomic imbalances, such as unemployment and inflation, as well as longer-run issues of development, productive potential and industrial restructuring. The harmonisation of such short and long-term aspects of planning is the subject of subsequent sections of this chapter.

A central focus for all planning exercises is investment. One of the reasons why it has been traditionally argued that investment should be determined at the national plan level has to do with the question of intertemporal resource allocation. In the political economy and left social democratic traditions, it has been argued that this cannot be left to the market for the balance between consumption and investment is a political question (Nove, 1983, p211). And thus if the national plan can control aggregate investment, 'society clarifies for itself the decision-making process by which it creates it own future' (Desai et al, 1988, p77).[1]

The importance of investment is also related to the present world economic conjuncture in which most countries feel a pressing need to adjust and restructure their economies.[2] In this context, planning is seen to be able to make a significant contribution both to raising the volume of investment and to improving its structure, as well as to related issues such as research and development, technical education and training. The task is both to improve the efficiency of resource allocation and to influence the quality of investments by lowering the incremental capital output ratio through the promotion of innovative schemes for process and product development, the reorganisation of the production process and the transfer and diffusion of technology (Hughes, 1986, p222).

Hare points to three main tasks for the national plan with respect to investment, which are common to other models of planning (Hare, 1985; Hughes, 1986; Gilhespy et al, 1986; Cowling, 1987). Firstly, an investment plan can be formulated, in part based on the desired (and realistic) employment creation and growth targets. This would point to the aggregate investment needed over the plan period which can be broken down by sectors of industry and other relevant categories (such as public and private sectors). Secondly, the proposed investment plans of the public sector and dominant private sector firms would have to be examined to determine bottlenecks, plan inconsistencies and the extent of resources needed to support investment which existing firms did not intend to undertake. Thirdly, individual investment project proposals could be assessed to ensure their financing and to support organisational changes for the formation of new firms, or mergers between existing ones (Hare, 1985, p222).

This leaves, to a certain extent, ambiguous the degree to which the above formulation can be seen in terms of the indicative planning model with the emphasis on co-ordination and mitigation of information problems to enhance plan consistency, or in terms of a more interventionist stance. The left social democratic model would support a degree of intervention that goes beyond indicative planning to ensure plan implementation. One of the consequences of the new attention paid by left social democracy to questions of production and the network of institutional and market relations which affect enterprise decisions is a new concern for what has been termed strategic planning (Best, 1986; Gilhespy et al, 1986; Murray, 1987).[3] Some theorists in the left social democratic tradition have been more influenced by the Japanese model of planning than the French. For instance Cowling (1987, p13) points out that the Japanese government goes beyond providing a suitable environment for industry or merely enhancing information flows within the economy. Its intervention in 'strategic industries' can form a basis for a more interventionist planning approach: 'the dominance of the regulatory function of the government has to be displaced by its developmental function - that is direct involvement in the birth, growth and death of industries' (Cowling, 1987, p13).

At the level of the enterprise strategic planning is concerned with such new developments in capitalist economies as 'post-Fordism' and the 'new competition'. It is beyond our scope here to go into the details, let alone assess the importance, of flexible specialisation, non-standardised products, the integration of planning and doing, and the need for firms to adopt strategies to a changing and shapeable environment. Crucial to this 'new competition' is that firms do not only compete on price but on a whole range of factors (marketing, product engineering, design, quality, service etc). The rationale for intervention here is not merely market failure but organisational underdevelopment - firms may lack the capacity to compete on one, or more, of these non-price factors (see Best, 1986; Gilhespy et al, 1986). As Desai et al (1988, p78) have argued the need for sectoral and enterprise planning reflects the fact that: 'rational economic systems do not simply respond to objectively existing market demand, but plan and innovate for the future. There are good reasons for believing that purely market-driven enterprises and their associated capital-market institutions tend to be myopic in their perception of the future'.

At the sectoral level the rationale for strategic planning is that different industrial sectors may face very different sets of problems, thus limiting the usefulness of policy at the macroeconomic or aggregate investment level. Furthermore, competitive market strategies may be mutually destructive. There may also be a need for influencing inter-firm organisation, or for sharing export promotion and R & D expenditure in a particular sector. As Best who has also been influenced by the Japanese experience has written, Japanese planning at this level seeks to foresee future sector product and process developments, providing guidelines for enterprise strategies and shaping the contours for industry. Furthermore, this: 'allows banks to evaluate investment proposals not as discrete projects but as parts of an integrated sector anticipated to become internationally competitive. The message to enterprises is clear: strategies must be consistent with projected sector development' (Best, 1986, p191; see also section 3.3.1 of this chapter). Strategic planning thus operates at both the sectoral and intersectoral level, by indicating those sectors which are viable in the medium to long-term given a framework of state intervention, and if needed, protection and new resources.[4]

It is clear that those influenced by such experiments in strategic planning would not seek any simple transfer of, say Japanese institutions, to their own country, clearly recognising that such institutions have to be adapted to fit specific political economy contexts. Thus both Best (1986) and Cowling (1987) argue, within the context of British debates about the appropriate form of planning, that strategic planning would have to be associated with institutions for seeking consensus within society and by extending participation over industrial strategy at various levels. Cowling (1987, p15) also notes that strategic planning could only be introduced gradually: 'a piecemeal/step by step approach would be appropriate to allow the growth of awareness of people about the relevance of economic planning to their own lives'.

It is clear that any development strategy specified in the plan is bound to be associated with certain costs. For instance a concern with restructuring and developing new industries implies that certain old plants or skills may become redundant. Furthermore this is likely to have uneven social or regional effects, damaging certain workers or regions more heavily than others. If the national plan is to have broad support within the whole of society then it must specify the policies that can mitigate an uneven distribution of costs. This may entail the plan ensuring resources for active retraining and placement schemes, that are typical in the Swedish economy, to preserve employment and enhance social and economic efficiency. The need to finance such schemes, as well as overall investment expenditures, clearly indicates the need for the national plan to be coordinated with macroeconomic policy to ensure the availability of resources (see section 3.5 to 3.7).

To conclude we have examined in broad terms the scope of the national plan. Such a plan in most cases will be the responsibility of a National Planning Agency, although its institutional co-ordination within the state has been left deliberately vague. It has also been suggested that a National Planning Council or similar institution will be necessary to act as a forum for debate on long-term planning for major industrial sectors and processes. Finally, left social democratic theorists have made clear that the national plan need not be comprehensive in a manner of the five-year plans of centrally planned economies. Clearly there are significant differences between supporters of planning on the degree of intervention or the rate at which it should be introduced. However even those on the interventionist end of the spectrum do not intend the plan to be rigid or comprehensive: 'in neither strategic planning of the future, nor allocative procedures of the present, will the system be a pure one. Industrial policy will partially supplant the market system, but the market will continue throughout the interstices of such strategic planning - an intimate blend of government strategy and the market' (Cowling, 1987, p13). That having been said the national plan framework gives the government the opportunity to go beyond forecasting and take a position on the desirable direction of the national economy. Within this framework it can foresee obstacles to such a plan and take corresponding action. However such an approach is based not only on the realism of the plan goals in general but also on a realistic assessment of the instruments available to ensure the implementation of plan objectives.[5]

3.2 Beyond indicative planning

Shonfield's (1965) planning typology delineates interventionist forms of planning from others by giving three broad categories. The first approach Shonfield calls intellectual/indicative and is based on the problem of uncertainty in a market economy (see chapter 2). The second relies on reinforced government powers: 'the state controls so large a part of the economy that a planner can, by intelligent manipulation of the new levers of public power, guide the remainder of the economy firmly towards any objective that the government chooses' (Shonfield, 1965, p231). The third method eschews wherever possible the use of direct government intervention and relies on a corporate formula for managing the economy. Interest groups are brought together for a series of bargaining sessions to help the economy onto a desired, and generally acceptable path.

Our concern in this section is with the second category, although as we shall see the various proposals rarely promote a 'pure' system, relying on elements from all three categories. This can be seen in a recent statement on socialist economic strategy which emphasises the need to promote both economic efficiency and modernisation as well as social goals: 'to ensure this, relations of economic power will have to be transformed through regulation, social ownership, economic democracy and state intervention'.[6] There are two points that need stressing here. Firstly, that the left social democratic 'alternative' rests not merely on different economic policies but on a political economy project concerned with transforming power relations within society. Secondly, that the recent trend has been to emphasize the need for a large number of mechanisms for extending public control rather than relying on a sole regulator such as 'national planning' or nationalisation (Hodgson, 1986; Best, 1986). Here we examine those mechanisms and institutions, which are intended to integrate the public sector, and influence the private sector, into the objectives of the national plan.

3.2.1 Public and social ownership

There are a wide number of reasons why socialists have traditionally argued for an extension of public ownership as an integral part of their strategy. Some of these can be rationalised by the orthodox conception of market failure (eg monopoly, externalities, social cost-benefit considerations). The need for restructuring, and the high cost or risk of certain investment projects has also often entailed a role for the state. Thus, the rationale of nationalisation has often been to promote capital accumulation, even in the long run, within the bounds of a capitalist economy (Murray, 1987, pp93-4).

However, as Murray points out, socialists have also been concerned with extending public ownership on grounds that may conflict with capital accumulation. The rationale here may be to reduce the power of private capital (through its control of investment) and promote a more diverse ownership structure within the economy, to increase the share of profits going to the public sector, to improve working conditions or extend participatory mechanisms within enterprises (see section 3.4).

The existence of such a diverse range of rationales for extending public ownership creates a number of problems for any alternative economic strategy. It is of paramount importance to clarify which set of arguments apply to which sectors or particular enterprises. If the overall goal is to increase social control then it will be necessary to delineate the extent to which this necessitates public ownership, or other forms of social ownership, as opposed to relying on more indirect intervention.

Nove's categorisation of the four main types of possible producers of goods and services in a socialist economy is a useful starting point to some of the central issues and in particular, the question of relating the form of social ownership to the type of intervention which may be necessary in any alternative economic strategy (Nove, 1983, pp198-206).

1. <u>State enterprises</u>: these should be centrally controlled and administered. Some of these will be operating under large economies of scale and/or have important external effects on the rest of the economy. They will be enterprises in which decision-making will have to be dominated by national/central considerations. In cases where such enterprises have a strong monopoly position, there could be a social supervision of their activities through, for instance, management being responsible to representatives of the state, users and workers. Economic power can be devolved through internal structures promoting operating divisions or area/regional boards (Gilhespy <u>et al</u>, 1986, p116). The degree to which such firms can be tightly regulated, but still decentralised and open to some degree of competition, can be investigated. However, the state would have final responsibility.

2. <u>Socialised enterprises</u>: these could be either state or socially owned with a management responsible to the workforce. These are likely to be working under a more competitive market framework with fewer external linkages to the rest of the economy. These firms are thus likely to need more autonomy and flexibility although they could still be influenced by state intervention by the type of strategic planning already discussed. The precise nature of socialised enterprises and how they are to be distinguished from state enterprises is not very clear in

Nove's analysis.[7]

3. <u>Co-operative enterprises</u>: here too firms will be operating under competitive conditions and their decisions on production, prices and investments would not influence to any significant extent the rest of the economy. In contrast to the socialised enterprises, co-operatives could freely dispose of their property and would be likely to be relatively small (Nove, 1983, p206). Experience has shown however, that state intervention in the form of support facilities (for starting up, product development, marketing, training and legal advice) is indispensable for the success of co-operatives (Gilhespy <u>et al</u>, 1986, p52, 126).

4. <u>Private enterprises</u>: here of course, by definition, any government intervention to enhance social control will be indirect through mechanisms such as financial incentives and enterprise boards' equity stakes, which will be discussed shortly.

Given this typology one can see that any strategy for social control has a number of options, depending in part on the rationale for intervention. Depending on the rationale in question, public ownership may necessitate full public ownership while in other cases a minority stake or merely indirect controls or regulation may be appropriate. Consider the case for public ownership which is based on the need for technology diffusion in a strategically important sector with the existence of significant externalities. In high technology or dynamic new sectors the state may need to extend public ownership, fearing that indirect controls may be insufficient. Here the state may help to set up a new public enterprise which would remain in a competitive environment. Such a firm could mitigate the effects of asymmetric information which often means that the state is not aware of the true demand and supply curves facing a particular industry. This would facilitate the state's strategic planning of the sector by providing it with more adequate information to plan the necessary financial resources for that sector. However this entails that the government works out in which sectors such arguments are relevant and also the sort of enterprise and/or level and extent of regulation which is necessary.

Gilhespy <u>et al</u> (1986, p116) have usefully delineated four criteria for facilitating decision-making in this area. Firstly, there is the question of ownership, the degree to which public ownership is needed and its form. Secondly, one should ask at what level intervention is necessary. For instance economies of scale arguments may necessitate intervention at the national level, while in certain industrial enterprises a regional or even local response may be more appropriate. They add that: 'to achieve the devolution of economic power, the minimum necessary level of intervention may need to be pursued' (Gilhespy et al, 1986, p116). Thirdly there is the issue of management structure where the type of workers' participation depends crucially on the nature of economic decision-making. As Nove (1983) has stressed the larger the production unit or the more there is a good case for taking decisions at a central level, the less is the scope for labour participation and self-management. There are various possibilities for extending participation: setting up co-operatives, promoting collective bargaining within private enterprises or setting up supervisory mechanisms for state firms (see section 3.4). The final criteria is that of industrial activity. Production technologies, economies of scale, and development costs vary, and this will also influence the level of state intervention and the form of social ownership required.

From the above it is clear that the rationale for planning, and extending social ownership is, in part, based on questions of industrial reconstruction and economic development. Even when this is the case, and long-run capital accumulation is promoted, private capital may oppose such measures on political grounds. However the opposition is likely to be even more vehement if the rationale is for some of the social goals of left social democracy (Murray, 1987, pp86-7). Such goals may include promoting industrial democracy, increasing equalities between different groups of workers and improving working conditions and the quality of work. Murray provides an interesting table which posits some likely effects on short and long-run market competitiveness of various goals (Table 3.1).

Murray then uses this table to show the extent to which public or socially owned firms will be under pressure not to implement a strategy which promotes some of these social goals. Firms will be under pressure both from the political hostility of the private sector and from the criteria of market profitability. Murray (1987, pp98-112) points out that such problems are at the root of the explanation of why nationalised industries have so often operated in a manner indistinguishable from that which would have been the case if they had been privately owned. This leads him to consider the question of how progressive

Table 3.1
Market Competitiveness and the Reasons for Social Ownership

Reasons for Social Ownership	Improves Market Competitiveness		Worsens Market Competitiveness Loss-making	
	Short run	Long run	Short run	Long run
Rationalization and restructuring		X	X	
Control of monopoly power	not applicable			
Macroeconomic planning and stabilization	nationally neutral			
Social costs-benefits			X	X
Long-run investment		X		
Improving wages and conditions, work processes, flexible working hours, training		?	X	?
Extending democratic control within the workplace		?	X	?
Increased equalities	?	X	?	
Providing Services for need			X	X
Ensuring variety and production for minority demands			X	X
Retaining plants and industries which are not market-viable, for social or strategic reasons			X	X

NB A question mark denotes areas where the effect of a particular goal is uncertain. For instance, we have seen that socialists have argued that participation, in the long run, may actually promote productivity and efficiency.
Source: R Murray (1987) p96

governments can ensure that state managers put its policies into practice. The question of the internal structure of enterprises is a crucial issue which is considerably undertheorised in the existing literature on planning. As Murray (1987, p96) concludes: 'the relationship between socially owned enterprise and the market must be at the centre of strategic political thinking about the public sector ... Publicly owned firms ... will find themselves operating 'in and against the market". Murray's discussion is mostly concerned with the relationship of social goals and social ownership. But many of the arguments would hold also, perhaps more acutely, when the goal is to extend these social aims to private sector firms.

There has indeed been a tendency for left social democrats to place a new emphasis on promoting diverging forms of social ownership. This may involve nationalisation or merely the state acquiring equity stakes in a firm; it may mean local enterprise boards intervening at the local level by setting up co-operatives or acquiring shares in particular firms. Our argument here has been to indicate that if the overall rationale of public control is both to diversify forms of ownership as a 'good in itself' and to promote economic efficiency and modernisation, then any strategy must clarify such issues as the rationale for public and social ownership, in which sectors/enterprises its relevance lies, its scope and at what level intervention is most appropriate (national, regional, local). This will also clarify where social control can be better operated by more indirect means. This is likely to be important given the limits imposed on extending public ownership by financial constraints, time consuming procedures of legislation and a whole set of problems associated with administrative and bureaucratic intervention. Finally such clarity can help formulate the 'rules of the game' of economic decision-making within society. Any extension of public ownership, whatever its form, is likely to be opposed by the private sector. However since left social democratic models give a significant scope to both private enterprise and the market, the 'rules of the game' must be clarified.

3.2.2 Enterprise boards and new public enterprises (NPEs)

We have argued that any left social democratic strategy is likely to go beyond indicative planning to intervene in the productive economy. Furthermore the argument is that if 'strategic' planning is to be effective then there is a need for mechanisms and institutions to ensure the implementation of plan objectives. Here we examine the role envisaged for enterprise boards and new public enterprises (NPEs). Both of these have featured, as a fundamental component, in a number of alternative planning proposals.

We begin with the planning proposals developed by the Labour Party in Britain in the 1970s.[8] The debate in Britain was deeply influenced by the theoretical work of Holland and his model of how a National Enterprise Board (NEB), new public enterprises and planning agreements could be integrated into an overall planning model for the British economy (planning agreements will be discussed in the next sub-section, 3.2.3). The emphasis was on the supply-side response of the economy, which we have suggested is a characteristic of the thinking of left social democrats in this period. Similar developments were taking place in French debates, around the 'Common Programme' of the left. Since Holland's model[9] was based on an assessment of the structural problems of the British economy, provided an extensive debate on the role of planning in socialist economic policy-making, promoted a radical package of planning mechanisms and institutions, and was influential in the development of Labour's more interventionist economic thinking of the 1970s and 1980s, it is a useful starting point. We look at Holland's proposals before examining some of the problems associated with them. This will allow us to go on to investigate more recent developments on the role envisaged for enterprise boards and NPEs.

Holland (1975) argued that in an economy with structural problems, indicative planning was of little use. Firms were unlikely to expand their activities if they feared this may leave them with spare capacity. Furthermore developments within the capitalist economies, such as the growth of multi-national corporations (MNCs) and industrial concentration, had severely limited the influence that could be exerted on such 'mesoeconomic' firms by traditional Keynesian mechanisms (for instance, interest and exchange rate manipulation and regional incentives).

Holland was influenced by the Italian experience with the State Holding Company, IRI. He noted that between 1968 and 1972 state companies, through planning agreements, were required to accelerate their investment programmes, playing an important part in restraining the recession - although the size of the public sector was too small to have a larger effect. Investment plans for one industry could be consciously related to the effects of plans of other industries - IRI could expand a new industry in a region where some declining industry was closing production (Nove, 1983, p174).

Thus Holland's conception was to 'harness' mesoeconomic power to plan the medium term and avoid the main waves of the stop-go cycle, which had so bedeviled planning experiments in the 1960s. This was to be achieved by setting up a National Enterprise Board (NEB) which was to control about one firm in each of the important sectors of manufacturing. The NEB had six objectives in the 1974 Labour Party manifesto:

1) stimulation of investment;
2) creation of employment in areas of high unemployment;
3) increasing exports and reducing dependence on imports;
4) promoting industrial efficiency;
5) countering private monopolies;
6) preventing British industry from passing into unacceptable foreign control.

The key to Holland's analysis is that although given 1) the other objectives are basically reinforcing, 1) could not come about by the mere application of demand management (Holland, 1975, p183). Firms might not expand since they are aware from past experience that an increase in government expansion during the 'go' part of the cycle may quickly be reversed. Furthermore Holland pointed to the increasing investment horizon necessary for large enterprise in modern industry which extended beyond stop-go budget cycles and the tendency for MNCs to invest abroad.

Holland referred to both 'push' and 'pull' factors to remedy this situation. The 'push' factor is secured by the new public enterprises (NPEs), controlled by the NEB: 'directly undertaking investment projects which the private sector leaders have hesitated to promote' (Holland, 1975, p184). Holland repeatedly stresses the need for this to be a broad (in many sectors) and significant wave of investments. The 'pull' effect can be understood in terms of traditional oligopoly economics. In microeconomic theory oligopoly/monopoly is inefficient exactly because firms have an interest in restricting output and raising prices. However if a public firm expands because it is interested as much in the social as in the private benefit (it can accept normal, as opposed to super-normal profits) then other firms will have to follow or risk losing their markets. This, of course, implies that leading manufacturing firms do in fact have super-normal profits, an empirical question beyond our scope here. It explains why Holland stressed the need for the NEB to be involved in profitable and efficient firms.

What can we say about Holland's conception of a planning alternative? Theorists were aware from the start of the potentially explosive political economy implications of the NPEs. For Holland argued that, in Britain at least, the NEB could not rely on existing nationalised industries which were concentrated in basic industry, economic infrastructure and social services. Many of the sectors involved were, in Holland's terminology, 'growth dependent', depending for their output growth on the active growth-initiating sectors of manufacturing and services. Neither 'push' nor 'pull' factors could work with such a base. The post-war consensus, however, had not envisaged the nationalisation of profitable companies (Blazyca, 1983).

Private capital is likely to oppose nationalisation or the use of NPEs on economic as well as political grounds. Hare (1985, p227) considers the case of setting up a NPE in a manufacturing sector where the planners judge there is a potential not currently being exploited by the private sector. He notes that if the planners have made a correct decision then this may not adversely affect the private sector, but if they are wrong the existing firms will face increased competition which may affect their profits. Nor would such an opposition necessarily be limited to private capital:

> the workforce in a company faced with bankruptcy because of competition from a public firm receiving subsidies or operating with lower rates of return will be just as vocal as its owners in opposing 'unfair' competition from the state. This is of course why one of the cardinal principles of the present structure of aid is 'non-discriminatory' between competing domestic companies. (LWG/CSE, 1980, p72).

Selective intervention is also associated with a set of problems such as the risk of ignoring other important areas in the economy or reducing the market constraints operating in favour of competition. Another issue is the ability of the state, or bureaucrats, to 'pick winners'. We can draw three conclusions from the above problems associated with the use of NPEs and selective intervention. Firstly, that such intervention is likely to affect the competitive environment and incentives faced by firms in a particular sector, entailing a careful assessment of the overall needs of the sector itself. This reemphasises the need for strategic planning at the sectoral level (section 3.2.1). Opposition by existing firms can be mitigated through participatory regulatory institutions which discuss sectoral priorities. However there is here the well-known problem of 'regulatory capture' whereby firms dominate the institution that is supposed to be regulating them: 'Hence regulatory capture both confers benefits and imposes costs. It has not proved easy anywhere to gain the benefits without incurring at least some of the costs.' (Hare, 1985, pp228-9; see also section 3.4 of this chapter).

Secondly, as Holland clearly accepted, his proposals entail a major 'change to the balance of power in society'. This entails, as a minimum, that there exists a strong support within society for such a planning exercise. It explains why many analysts have suggested the need for a gradual approach which would enable potential supporters to see the benefits of planning (Cowling, 1987).

Thirdly, a gradual perspective is also indicated by some of the problems associated with the state, or a NEB, 'picking winners'. On the one hand, there may be strong political economy grounds for the state undertaking such interventions. As Cowling has pointed out, the question is not merely one of the efficiency of economic decision-making - that is to say whether the private sector is more suited to making economic decisions than 'bureaucrats' or planners:

> the essential difference between the public and private selection process lies in the difference in perspective - in the ultimate objectives of such a process rather than the efficiency with which any objective is pursued. The aim of government in the area of economic policy is to create a dynamic and productive national economy and private decision-making may not be consistent with this. (Cowling, 1987, p14).

Given this goal of creating 'a dynamic and productive national economy' and the overall approach to economic policy adopted by left social democracy, any alternative economic strategy is likely to attempt, to some extent at least, to 'pick winners' - to target support at specific sectors or firms. As Geroski (1988, p22) argues:

> In principle, the choice between sectors that one might consider supporting hinges on identifying market failures, and, in practice, most lists of potential 'winners' are generally pretty similar. The difficult questions are those concerned with how to go about 'creating winners' from what are merely possibilities that may fail to materialize if left on their own. General, non-discretionary assistance has the sole virtue of being (relatively) administratively simple to provide, but this is achieved at the cost of providing support that is not necessaily well-tailored to the needs of most recipients, that can (in the case of subsidies) undermine the incentives of most recipients and that provides suppport for (perhaps many) firms that do not need it. Although it is by no means certain, it is hard to believe that more discretionary, better-targeted, and more carefully custom-designed policies cannot provide more satisfactory results for the same input.

On the other hand, there are clearly certain problems with respect to the state's <u>capacity</u> to 'pick winners' which need to be examined. It is to these we now turn by considering the problem of control. For Holland's conception seemed to consist in the idea that the experience of state holding companies, and planning agreements, in Europe could be transferred to Britain, but on a much larger scale. However there exists a <u>prima facie</u> case for thinking that such an expansion would create new problems of control in such a quantitative manner that there would be a qualitatively new problem.[10]

There was, for instance, little discussion concerning how NPEs, under the auspices of the NEB, were to be controlled - will these firms not have their own inclinations, strategies, and interests? True part of the answer lay in the planning agreements system which we have yet to discuss. But the emphasis on the role of NPEs left open a number of crucial issues pertaining both to the internal structure of those firms and the workings of the NEB itself.

The debate on the socialist countries seems to suggest that the state's capacity to intervene is not boundless. As Murray has pointed out, quoting Tawney, even if the state does represent the interests of the community, 'when the question of ownership has been settled the question of administration remains for solution' (Murray, 1987, p88). At stake is the state's 'capacity effectively to subordinate all the enormous number and variety of state agencies to its control' (Tomlinson, 1982, p51). The mere ability to make laws does not give the government or parliament such control. The ability of the government to co-ordinate its various ministries and nationalised industries has often proved very difficult. We saw such problems in chapter two when discussing the French government's inability to integrate the public sector into its planning. It is probable with the type of planning being discussed here that such problems are likely to be more acute. The state will not only have to set up new institutions, but also integrate them into the existing system of administration. Tomlinson (1982, p144) is particularly critical of Holland's approach to NPEs: 'the objectives are then multiple, paralleling the multiple obstruction placed in socialists' path by mesoeconomic firms. This framework retains the notion of the sovereign firm, possessor of economic power. As a result, a heterogeneous range of socialist objectives are expected to be accomplished by the relative simple expedient of confiscating that power'.

Desai et al (1988) have argued that the underestimation of the importance of such questions in the planning literature rests on a misunderstanding of certain developments within capitalist economies. They point out that a number of left strategies for the introduction of planning have been based on the premise that planning has been an ever-growing trend within capitalist economies and in the internal operation of capitalist firms. While not doubting that there is something to be said for this view, they doubt its relevance to alternative models of planning. Firstly they question whether the growth of planning can be seen as a linear, non-reversible process in capitalist economies. They point out that many marxists have not been appreciative enough of the extent to which class struggle exists both in the relations and forces of production. The latter process entails that the importance of markets can contract or expand, according to the balance of class forces, rather than planning having an inexorable growth due to the 'technical' requirements of the forces of production. Secondly, that the planning actually carried out by capitalist firms is only understandable in the context of market alternatives, that is the existence of the market 'existing outside'. The consequence of this is that the organisation and direction of the firm has not become a purely 'technical problem of administration'. Such arguments caution against the political voluntarism implicit in certain strategies based on taking over certain enterprises and merely instructing them to carry out different goals.

We have noted in the previous section that the planning alternatives we are examining are predicated on the continuing existence of markets 'existing outside'. Thus there is a pressing need for a more detailed analysis of the relationship between planners and the NPEs than has so far been evident in the theoretical literature. The relationship between planners and firms is an extremely complex one and as Hare (1985) has written, in Britain, even during the last war, where matters were considerably simplified by the concentration on a relatively few number of objectives one should accept that: 'the practical difficulty of interpreting an overall objective of an individual firm and its production possibilities is immense: it is extremely unclear what exactly it requires the firm to do' (Hare, 1985, p27). Part of the solution to the problem will necessarily rest on a costly trial and error process. But what is also indicated is more work on the organisational structure of NPEs and the use of a greater range of regulatory mechanisms to influence the incentive structure that all firms, public and private, face within a particular sector.

One proposal which has shown an awareness of some of the problems being discussed here has come from Gilhespy et al (1986). This group favour an institutional framework which would replace the one state holding company model, with a number of smaller enterprise boards responsible for particular sectors. For instance, they propose setting up a 'British Enterprise' in telecommunications and one for motor manufacturing, with each enterprise board involving itself in the important strategic issues of the sector. This is in part to simplify the administrative problems that would face any large state holding company that would have to co-ordinate government policies in a number of sectors, involving numerous ministries, and thus entailing severe problems of information flows and co-ordination. However it would also facilitate, at a decentralised level, decisions on what new NPEs were required, whether only a small equity state in particular firms was necessary and how the policy of the whole sector was to be influenced by other mechanisms (see section 3.2.3). Such an approach could be initiated on the 'step-by-step' basis, by concentrating on priority sectors, thus reducing the administrative, personnel and management problems which would be associated with Holland's NEB, which would necessitate building up a large complex institution almost from scratch and relatively quickly.

Another response in the direction of reducing the comprehensiveness of the alternative planning institutions has come from the new emphasis on local and regional enterprise boards. We have already pointed to the need of delineating the level at which intervention is necessary. Much of the debate, in Britain at least, has stemmed from the practical experience of planning at the local level gained from such organisations as the Greater London and West Midlands Enterprise Boards (see Boddy and Fudge, 1984). Two arguments will be developed here. Firstly, that there is a need for a more decentralised model but that this should be seen as a supplement, rather than an alternative to the national planning institutions which we have examined. Secondly that the experience of local planning has suggested insights and feedbacks for national planning.

Hughes (1986) has suggested three reasons why there is a need for decentralisation. Firstly he refers to empirical work which suggests that local knowledge and information is important in the successful development of industrial/financial links. Secondly he notes that the recent British experience at the local level 'suggests that effective and efficient investment funding requires a close involvement in the details of the local economy and in the structure and organisation of the industrial enterprise being funded, as well as the setting up of quite detailed monitoring and reporting frameworks for companies which receive financial support' (Hughes, 1986, p226). Thirdly there exists a need to encourage participation to legitimise the planning exercise.

Local enterprise boards (LEBs) have both provided back-up support for local enterprise such as finance and accounts, and used this as part of a local economic strategy which emphasised, for instance, employment creation or insisted on good industrial relations. Hare (1985, p205) notes their role in revamping ineffective managers and management structures which could give more confidence for existing private sector financial institutions to invest (see also Gilhespy et al, 1986, p62). Much of the aid that has been channelled into local enterprise has taken the form of share capital, and worked towards 'socialising' the local economy (see section 3.2.3).

If Holland's NEB was influenced by IRI, then LEBs have been more of a development of GEPI (the Italian State holding company responsible for intervention in medium-sized and smaller firms) and the British experience with the Industrial Organisation Corporation (IOC). The role of LEBs is predicated on the development of local planning, the creation of local economic information and the ability to examine possible complementarities and to undertake branch surveys to examine possible firms to help. Hare also points to a more active role:

> as a result of investigating their own local economies, the enterprise boards should be able to identify some areas or branches where investments could be usefully stimulated. In other words, boards could move into the business of identifying possible projects for themselves rather than just wait for external proposals. Having identified some projects, it would be necessary to find, or even create, suitable firms to serve as vehicles for their implementation. (Hare, 1985, p206).

Most analysts have stressed that LEBs need a framework of state intervention in order to prosper (see Hare, 1985; Gilhespy et al, 1986). The state is vital in setting the general framework of development - in Britain in the early 1980s, LEBs have only had a small effect on employment given the rather hostile macroeconomic climate they have faced. Indeed some employment creation may have entailed job substitution rather than net employment gains. Nor can the concept of 'socially useful production' be seen as an alternative to state intervention. As Hare (1985, p162), amongst others, has pointed out either the products will be sold to the public sector in which case the viability depends on the structure and level of public expenditure, or if it is to be sold to the private sector, 'socially useful production' proposals come down to a critique of market research already being carried out. If again a subsidy is required then this too is a matter of state intervention.

Here we have been discussing the role that NPEs and NEBs could play as part of a planning alternative. We have also seen some limitations connected with Holland's original model. We have suggested that adherents of interventionist planning, while continuing to see the importance of the previous schema, have been emphasising a less comprehensive and decentralised approach. We return to an overall assessment of the various proposals in section 3.5.

3.2.3 Planning agreements, financial incentives and commercial leverage

We now turn to other mechanisms/institutions which have been seen to complement the approach being discussed here. Once more we follow the same procedure. We begin by examining the role played by planning agreements in Holland's model. We go on to appraise critically his approach before examining more recent developments on ways to integrate the public and private sectors into a planning framework. Central to Holland's conception of a 'new mode of planning' was his proposed planning agreements system. European experience was again important in the development of his thinking - the French experience with 'contrats de programme' and 'contrats de stabilite' as well as similar mechanisms used in Belgium and Italy (see Holland, 1975). Planning agreements would co-ordinate mesoeconomic NPEs and private firms as well as existing nationalised industries. The implication was that: 'the new mesoeconomic accelerator -coordinated through the Planning Agreements system - would ensure that changes in the aggregate level of demand could be matched by appropriate changes in the level of investment where this counts most - in the giant firms in the mesoeconomic sector' (Holland, 1975, p243).

Planning agreements had a large number of objectives which in the Labour Party's 1973 Programme included: to gather information about firms' past and projected behaviour, to use this information to clarify planning objectives, to reach agreement with firms on particular objectives, to provide a basis for channelling selective government assistance, to increase accountability in firms' behaviour. The institutional setting envisaged by Holland was that a Cabinet committee, responsible for economic policy, would involve the principal ministers concerned in industries and services brought into the planning agreements system. Below this the industrial co-ordination of the system could be undertaken by a planning division of the Department of Industry which would be responsible to the

Cabinet committee. It would require both the NEB and large private firms to submit planning agreements drafts.[11]

Holland emphasised the distinction between tactics and strategy. Since this is a central question to which we shall return in subsequent chapters it is worth quoting him in full:

> the government department concerned should not itself be drawing up the programmes of the companies and corporations which will be scrutinised under the Planning Agreements system. For both the existing and new public sectors, plus the other firms brought into the system, company management (whether conventional or worker controlled) would be left freely to initiate its own programmes. Government officials (and occasionally ministers) would then determine whether or not these programmes conformed with economic and social objectives. These are the strategic issues concerning the kind of information made available to it through the Planning Agreements. (Holland, 1975, p230).

Some firm draft plans would be in line with the state's overall perspective and some would not. However, Holland did not envisage the signing of an actual planning agreement as compulsory: 'the Planning Agreements system can be less than wholly operative, but more than indicative... the enterprise would show the government where it considers that divergences of public and private arise' (Holland, 1975, p231). Even if such divergences existed much useful information could be gathered, for instance as an early warning system of both closures and new investments and to clarify the areas where the NEB and PA mechanisms would have to respond.

What leverage can be brought on firms to carry out policies they would not have otherwise carried out? Holland mentions three types: the use of NPEs to fulfil a planning objective if this was resisted by a private enterprise leader; withholding government financial assistance from twelve months after the Planning Agreement system came into operation and the renegotiation of such assistance on the basis of revealed need; and finally the use of public and social audits to help put pressure on the firms. Holland concludes: 'the Planning Agreements system should operate as a synthesised bargaining process between the government and the giant private and public corporations which have become so important to the national economy' (Holland, 1975, p231). Holland also argued that by extending democracy the danger of corporatist solutions could be avoided.[12]

Does the above system of planning agreements provide a coherent response to the problem of influencing firms, especially in the private sector?[13] The problem of the use of financial incentives to encourage participation in a planning schema is a notoriously difficult one. For instance how does one ensure that private sector firms do not receive aid for projects they would have undertaken anyway, in effect turning the incentive into an award?[14] An associated danger is encouraging firms to maximise available grants and incentives - the effect of which is known from the literature of firm organisation in socialist countries.

In Holland's schema, planning agreements were predicated on strong financial incentives for firms to secure their participation. However the 1974-79 Labour government introduced only a watered-down version of this approach with no mechanism for monitoring or securing implementation, nor any penalties for any failure to reach the targets of any agreement. There is a consensus that such an approach is of very limited use (see Gilhespy et al, 1986, p101; Hare, 1985, p237).

Another obstacle to the planning agreements schema is the degree to which firms are able to control their environment over the medium term. If they do not it would be difficult to pool their corporate plans and expectations. Estrin and Holmes (1983) have argued that the empirical data does not back Galbraith's (1967) view that firms have detailed corporate plans extending over five-year periods which are capable of being implemented in practice. Rather these firms operate with a number of possible scenarios, not worked out in detail, to enhance flexibility, and those plans which are of operational significance are shorter-term financial plans.

There have been a number of responses to the above problems. With respect to the last point, one could argue that the whole point of a planning system is to provide a stable enough framework to allow firms to plan on a longer-term basis. As Hare (1985) has written the raison d'etre of such an exercise rests on the benefit, in terms of economic co-ordination or increased investment, that can result from firms adopting more conditional precommitments than they would do otherwise. Hare notes that this entails such commitments from both sides of those signing such a planning agreement: 'the example of the recent, highly successful Chrysler rescue in the United States, in which successive tranches of the agreed financial support package were made conditional on specified performance indicators, illustrates what can be achieved within a more tightly defined framework' (Hare, 1985, p238). This approach has two further implications. Firstly that if the government is to be able to keep its side of various agreements, then it is clearly of paramount importance that it is in a position to control its own macroeconomic policy (see section 3.6-3.8). Secondly that the state needs to monitor the progress made by firms to a much greater degree than is implied by Holland's distinction between tactics and strategy.

This still leaves open the question of the degree to which planning agreements can only be of use when firms rely on state financial incentives. One response has been that the problem is not of financial incentives as such but of coordinating the multiple instruments that the state has at its disposal in providing leverage over firms. Thus Hughes argues that in the 1974-79 Labour government, negotiated agreements on development were of great importance: 'In fact a very substantial amount of relevant planning experience was gained at the level of bargaining between the manufacturing enterprise and state 'planners' and in the large scale use of selective planning criteria' (Hughes, 1983, p47). Hughes mentions three such planning schemes (accelerated projects, sectoral schemes and selective investment schemes) and notes that these could have been much more effective if tax relief for companies during the crisis was given on a conditional basis. The problem for Hughes was that 'the elements for large scale planning were not put together' (Hughes, 1983, p49).

This principle of co-ordination of policy has been termed 'commercial leverage' by Gilhespy et al (1986, p102): 'little leverage will be obtained if companies can get finance and other (non-conditional) forms of support elsewhere. Unconditional assistance from government departments should be progressively or selectively reduced'. Their approach has been influenced by the experience of LEBs in Britain. They define commercial leverage as: '... leverage through packages of support tailored to the needs and opportunities of particular firms - including access to finance and markets, the provision of management consultancy and advice, support for training and research...' There are a number of features which distinguish this approach from Holland's conception. Firstly, planning agreements are not seen as separate from the NEB. State holding companies could use planning agreements as one of many mechanisms (equity stakes, setting up institutions for research and development, marketing or technology networks etc). This implies a more sectoral strategy for: 'It is only within the context of broader sector (process-wide strategies) that enterprise decisions can be evaluated' (Gilhespy et al, 1986, p101). Secondly, planners would need a greater knowledge and involvement in enterprises than envisaged by Holland's distinction between tactics and strategy. Thirdly, planning at the level of the enterprise could be enhanced by making support conditional on a business plan being drawn up in a firm, negotiated by management and unions.

The problems associated with such a coordinated response should not be underestimated. Co-ordination of much information and many functions, currently in most administrations dispersed through various ministries and at various levels would be needed. It may also require a planning of public purchasing. Murray has proposed the creation of a new department of finance and purchasing with the task of 'consolidating and making use of these powers within a central strategic plan'. He adds that 'the task - particularly on the purchasing side - would be a long one, but not impossible given modern computer technology. It is difficult enough to find out about purchasing, let alone control it in a local authority, quite apart from doing it across departments, public industries, the Health Service, local authorities, the armed forces, schools, universities, and so on' (Murray, 1977, p108). Once more the problem of control is a critical one.

3.3 A complementary framework for planning

Here we consider the institutional framework necessary for the introduction of some of the planning mechanisms and institutions we have been discussing. This section does not seek to be comprehensive since the appropriate institutional framework is likely to be highly specific to particular countries.[15]

3.3.1 Financial institutions

There is widespread agreement in the recent planning literature for some form of National Investment Bank (NIB). One of the most prevalent of market failures is the availability of long-run high risk funding for various projects (see Hughes, 1986). There are reasons to think that this problem is more acute in Britain than other developed countries where national equivalents of NIBs exist. Some form of control over the flow of funds would be an important instrument of planning and could contribute to the exercise of leverage over firms. Furthermore, financial institutions have a powerful control over the direction of the economy both in relation to government debt and in respect to important variables of economic policy such as exchange rates and interest rates. If there is general agreement on the need to extend socialisation in this area, there are still important disagreements on the extent and means.

Hughes notes that nearly all major industrial economies have a publicly owned or influenced financial intermediary to help create the appropriate finances for longer-term investments: 'Given government guarantees on market borrowing and a high degree of expertise in portfolio management and project assessment it should be possible for the bank to act as intermediary between public and financial resources and private industry, as well as to be a vehicle for the funding of state-run or supported projects' (Hughes, 1986, pp226-9). With respect to our planning schema, the NIB would be an important supplementary institution for both the NEBs and LEBs. It is thus linked to strategic planning: 'the NIB would be a source of industrial finance - through packages negotiated by British enterprise and local enterprise boards - in the form of both loan and equity capital ... In addition, both the Bank and the boards could provide advice on technology and management techniques. Finance should be tied to the provision - and acceptance - of these services' (Gilhespy et al, 1986, p131). Hughes, on similar lines, has suggested that the activities of LEBs/NEBs could be included within one of the operating divisions of the NIB. He also points to the holding-company function where the provision of equity takes the form of direct state ownership (Hughes, 1986, p228).

But Hughes and Hare have also forcefully argued that since providing finance may include subsidies for wider social considerations, then the planning agency should ensure that issues of subsidisation are transparent: 'It would be difficult for the proposed NIB to operate in other than a broadly commercial manner, applying substantially the same criteria to all projects that came before it.... it is essential that investment subsidies should be separately identifiable, rather than being lumped together into general investment funding' (Hare, 1985, p232). This would enable the NIB to finance the most socially useful projects, and enable it to rank priorities, an indispensable prerequisite given limited resources.

There is also widespread agreement that access to such NIB funds should be linked to the objectives of the national plan. In the French case, for instance, the Credit National required the sanction of the planning agency before making loans. However there is less agreement on the extent to which social control should be extended into the financial sector by nationalising commercial or investment banks. Some have argued that nationalisation should be limited, relying on the indirect mechanism of requiring banks to negotiate with the planning agency on criteria for allocation of funds with perhaps public ownership as a final sanction if this does not work (Gilhespy et al, 1986; Hare, 1985). Minns (1982), however, wants to go further and take the merchant banks into full public ownership. To an extent this question revolves around whether indirect controls are viable which will depend on the specific case and the political economy of the country in question.

3.3.2 Trade and capital flows

Given the increasing internationalisation of financial markets and the growing interdependence of the world economy, trade and capital flows may constitute a major problem for any project of introducing planning into a market economy.

An example of the type of policies that have been suggested for increased regulation in this area was the Labour party's proposals for changing the tax incentives faced by institutions, such as pension funds and insurance companies, investing abroad. The increased funds could then be available to purchase British government securities to be placed under the auspices of the NIB.[16]

However this may be only one aspect of the problem. Any alternative economic strategy is likely to confront, sooner or later, the problem of capital flight. For example, Kindleberger (1987) discusses what he terms 'a middle class strike' against the Front Populaire in France in 1936.[17] This capital flight may be propelled either by a speculative run against the currency or, more generally, by the fear of wealth owners of increased taxation (or in the extreme case expropriation). Given the highly charged political atmosphere, that usually accompanies the election of a government committed to an alternative economic strategy, capital flight can occur even if the new government has made clear its determination not to increase taxes on wealth.[18]

In this context it is not clear that measures to alter the tax incentives which asset holders (including financial institutions) face will be particularly effective. Any tax penalties would almost definitely be outweighed by speculative gains if the currency were to depreciate or by gains in terms of a perceived reduction in risk. Thus there may also be a need for some form of exchange controls. Indeed as Grahl and Rowthorn (1986) noted about Labour's proposals, discussed above:

..what is worrying, however, is that Labour leaders seem to have accepted the fabulous argument that exchange controls are no longer possible now that we live in an electronic age. As a result they will have no means whatsoever of shielding the British economy from the pressure of financial speculation, and their entire programme will be at the mercy of international finance ..... Because exchange controls would not be perfect, this does not mean that they would be useless.

It is, however, notoriously difficult to be specific about the appropriate form that exchange controls should take and the proposals are likely to be country-specific.

To what extent should control be extended to multi-national corporations (MNCs)? Some have argued that the planning alternative should not be too concerned with this issue - Hare (1985) suggests that MNCs have not been responsible for Britain's economic decline and thus planning can ignore them. However others have argued that MNCs have an important effect on national economies and that therefore control over trade and capital flows is vital for: 'the international allocation of production, investment and jobs will be determined by distributional considerations, and although privately efficient for the transnationals, these may have no relation to a socially optimal allocation' (Cowling, 1987, p12). Here too the obstacles are numerous. Thus nationalisation of a subsidiary of a foreign MNC may be a blunt instrument, if that subsidiary is only a minor part of the overall circuit of production and distribution of the MNC. Murray (1981) writes that in such cases the expansion of a British-based firm, or the setting up of a new firm to challenge the MNC may be more appropriate.

The government has a number of instruments at its disposal such as regulating MNCs or denying them access to domestic markets. Such policies, it is now generally recognised, are more potent at a supranational level, for instance through the EC. This is both in respect to access to markets and reducing the ability of MNCs to play one country against another, and in setting up international regulatory codes for the operation of MNCs (see Cowling and Sugden, 1987).

Most governments have strong interventionist mechanisms, such as export trading companies and export insurance organisations, for supporting national industries. These are usually joint ventures with public dominance, and operate like sectoral firms for international trade, providing information for markets and potential markets, organising international trade flows, and therefore reducing information costs for firms. These too can play some part in creating a favourable framework for development within the context of an alternative economic strategy.

3.3.3 Competition policy

The overall framework in which firms work will be influenced by the competition policy operated by the government and there are good reasons to think that this too should be integrated into the concerns of planning. Thus over mergers policy, the emphasis would presumably be on investment, innovation and restructuring. In this respect Hughes points to the need for criteria that go beyond the effects on competition to include the central objectives of the planners (Hughes, 1986, p225). He also proposes that restrictive practices legislation should be flexible and expeditious to encourage the type of inter-firm collaboration that planners may wish to promote.

Competition policy would also have to avoid monopoly situations. The issue of pricing policy and quality standards is a complex one, and whatever strategy is adopted, it will have to take into account the needs of firms with respect to research and development, investment, etc. Przeworski (1985, p215) has pointed to the importance of using the taxation of firms to encourage investment by, for instance, combining high nominal tax rates on profits with high marginal rates of investment tax relief.

One final aspect is worth considering here. One of the limitations of how alternative planning has be conceived is that it has been premised on the overwhelming importance of manufacturing industry. In fact there is a trend within modern capitalism for a partial shift of power towards the retailing sector, with large retailing chains controlling, by various methods such as subcontracting or quality controls, independent producers (see Murray, 1987; Gilhespy et al, pp138-140). Here some of the planning powers that have been discussed in this chapter may be of little use. This necessitates new thinking on how the retailing sector can be influenced through various strategies such as promoting the access of small producers to the market as a countervailing force to large retailers, fashion centres for local clothing firms that could not provide for themselves, marketing boards, etc (see Gilhespy et al, 1986, p139).

3.4 Industrial democracy and planning

The type of planning we have been discussing so far has been concerned mainly with questions of economic efficiency, development and restructuring. But of course these aims are not in themselves socialist. It is for this reason that planning has often been undertaken in capitalist economies by governments that would scarcely describe themselves as socialist. Conversely we have seen the influence of the Japanese and French models on the type of planning mechanisms and institutions supported by left social democrats. But it is clear that socialists, of all persuasions, have a wider range of social and economic goals in pursuing alternative economic policies.

A classic example is the desire to extend industrial democracy, to intervene in the social relations of production. It is with this question that we are chiefly concerned here, although there are a number of other social goals which could be examined, such as the protection of the environment and income redistribution.

In particular we shall examine the relationship between industrial democracy and planning. Should these be seen as separate issues or does the fate of one depend on the implementation of the other? There is considerable disagreement in the theoretical literature over this question. Thus Hare (1985, p158, p164) argues that industrial democracy should be supported as a 'good in itself', to the extent that it does not contradict other goals, and its fate should not be tied to that of planning. In short, industrial democracy is an end, not a means (see also Burchell and Tomlinson, 1982).

Others have argued that an indispensable corollary of planning is the introduction of industrial democracy. Thus some have argued that the planning agreement schema, discussed earlier, was fatally flawed by the fact that it was not linked to a system of industrial democracy. Without information rights or forms of collective bargaining over enterprise decisions, there was no countervailing force within enterprises. The result is that planners discuss planning agreements on the terms set by private enterprises, thereby reducing planning at best to a form of corporatist bargaining between the state and private capital, while ignoring the interests of workers and the wider social goals of the community (see Wilson and Green, 1982, 1983; TUC/Labour Party Liaison Committee, 1982).

We shall not attempt to adjudicate between such conflicting claims but merely examine some of the central questions that any left social democratic party achieving power will have to confront. One way out of the dilemma is what we can term the dual rationale approach. Thus it could be argued that industrial democracy should be expanded, as a good in itself for promoting social control, but this should be done to the degree that it encourages reconstruction, economic development and efficiency. This has the added advantage that, to the degree that it works, it will widen the support for the idea of industrial democracy. The viability of this approach rests, as we shall see, on a number of factors.

At the level of the enterprise some have argued that far from there being an inherent contradiction between participation and economic efficiency, the empirical evidence shows that participation actually enhances efficiency (Hodgson, 1984). Such observations are partially based on the extent to which modern capitalist enterprises need consultative devices, in part because of the knowledge held by workers in the areas of technology and product processes. Even if such consultative processes are strictly limited at present, there is no reason why expanded participation and enterprise planning could not be adopted for the different and wider social goals for the enterprise supported by socialists (Murray, 1987). There is widespread support for such planning at the enterprise level, to encourage plan proposals from workers and to expand collective bargaining to questions of new investment projects or plant closures (see Hare, 1985, p156; Gilhespy et al, 1986, pp106-7). With respect to the conditionality of government finance, some have argued that state aid should be conditional on firms negotiating an enterprise plan with their workers (see Cowling and Sugden, 1987, p154; Gilhespy et al, 1986, p109).[19]

Similar arguments hold for the extension of participation at the sectoral, and intersectoral levels. Given the importance of planning at the sectoral level, it is clear that workers must have some say at this level which will also have positive feedbacks into industrial democracy at the enterprise level: 'BL [British Leyland] unions, for example, may negotiate a business plan with the company, the viability of which would be threatened if the government encouraged a Japanese manufacture to expand; conversely the sectoral rationale for running down Vauxhall/GM would have had no authority unless unions had been involved in drawing up the strategy' (Gilhespy et al, 1986, p107). If the rationale for planning is to take decisions which are socially optimal, which the existing structure of firms and their decision-making do not take into account, then the input from the 'community' interest must somehow be institutionalised.

There are, however, a number of reasons to think that the 'ideal' solution of the dual rationale approach, where economic efficiency is compatible with social goals, is unlikely to be without significant problems and conflicts of interest. From the literature on planning in centrally planned economies, and the self-management model of Yugoslavia, it is clear that there are a number of possible conflicts between a self-managed enterprise and central planners. Under various assumptions it can be shown that such enterprises may prefer increased wages to investment, or wish to restrict employment expansion (Nove, 1983). Such problems do not provide any reason for not extending participation at the enterprise level to allow workers access to information, the ability to propose alternative plans and negotiate over business plans. The question becomes more difficult over the extent to which workers can have a decisive influence over investment policy (see Hare, 1985, p156-7).

It is important when discussing the issue of industrial democracy to be clear on the way that property relations, as well as the level of desirable state intervention, may impinge on the degree to which it is feasible to extend participatory mechanisms. In private sector firms the management will certainly have to be responsible to shareholders, and to that extent workers' say in investment decisions is bound to be limited. Of course, as Hare again notes, to some degree shareholders, workers and managers may have a basis for consensus, since they all share an interest in the long-term prospects of the firm. He concludes that while conflict is bound to arise: 'the really essential question is not whether there would be disputes between shareholders' and workers' representatives, on the board, but whether they would be more frequent and intractable with a system of workplace democracy than otherwise. Personally I doubt whether they would be' (Hare, 1985, p158). In socially or publicly-owned firms the workers' influence over investment decisions can vary. We argued in section 3.1 that the left has recently been arguing for a greater variety of social ownership and to that extent the role of workers participation can vary according to the nature of a particular enterprise, its size, the level of externalities, the level at which state intervention is necessary (national/local), technology etc.

A central problem is how to promote industrial democracy, as a concomitant of planning, while at the same time avoiding the dangers of 'enterprise syndicalism'. As Burchell and Tomlinson (1982) argue, planning agreements:

> would very probably stimulate the production of workers' plans and social audits which seem to us a perfectly reasonable and intelligent approach for employees to adopt when negotiating corporate strategies. Furthermore it seems totally unrealistic not to expect that those who have been actively engaged in forging a strategy for their enterprise not to become attached to it and prioritise it over strategies determined elsewhere.

Such problems may be attenuated by participation at sectoral or inter-sectoral levels. But there is a danger of tripartism being the be-all and end-all of economic strategy. Hall (1986, p91) has noted how the 1960s British experience of tripartism in particular sectors helped to perpetuate rather than eliminate structural defects. Such institutions as sectoral working parties face the danger of leaving 'the incentive for rationalisation up to private sector actors themselves, and, faced with a relatively unchanged set of market incentives, they clung to traditional practices'.

There are thus a number of problems associated with integrating industrial democracy and planning. No easy resolution exists but we can perhaps state a few tentative conclusions. Firstly that industrial democracy is not an alternative to planning. Thus even in a hypothetical pure self-management model, the case for intervention by planners is strong, for the reasons we have developed (Nove, 1983, pp223-4). In short, the interests of a particular group of workers may be in conflict with that of the wider community. Furthermore, the upgrading of industrial democracy at all levels will be a difficult and time-consuming operation. In hardly any country are trade unions adequately prepared, ideologically or organisationally, for such a new role. It seems likely therefore that the institutions and mechanisms of planning can be introduced at a faster pace than can industrial democracy. Finally given the political opposition to industrial democracy, the state may have to intervene actively to promote it through its powers of social ownership, commercial leverage and so on.

3.5 The coherence of the planning alternative

In previous sections we have examined the various components of a possible left social democratic planning alternative. It remains to consider the overall coherence of such an alternative, and to point to areas of inconsistency or ambiguity within the overall model.

Our previous analysis should have made clear the package nature of many of the proposals. For instance the creation of a NIB is linked to the other elements of the planning framework. Thus we have seen how it is intended to work with both the NEB and LEBs to provide a package of measures (finance, consultancy etc) to help investment projects. Some of the loans could be in the form of share capital to extend socialisation of the economy. However if the response is to be coherent then the operation of the NIB has to be integrated into the objectives of the National Planning Council.

Cowling and Sugden (1987) and Hare (1985) have argued that such a planning package should not seek to be too comprehensive or centralised. They foresee a loose hierarchy between planners, industrial ministers and the NIB, with much of the work being decentralised to sectoral agencies and regional and local enterprise boards: 'The DTI [Department of Trade and Industry] would remain as the central coordinating agency but it would lie at the apex of an information structure, not a command structure. Variety and flexibility would be the essence of the planning structure and there would be no necessary attempt to cover all sectors, nor all regions, of the economy. The areas of activity would reflect the priorities established at the centre and within regions' (Cowling and Sugden, 1987, p154). We have suggested a number of reasons for a less comprehensive and decentralised approach, including the importance of delineating the level at which intervention is necessary and, indeed, possible, as well as the state's capacity to control its various public enterprises and holding companies. A further consideration is the opposition which will confront the alternative planning package. A 'step-by-step' approach, concentrating on priority sectors, if successful, may maximise the potential support within society for planning. This also applies to the question of extending participation, and industrial democracy, within the planning framework. Such participation requires considerable resources (trade union and community economic and technical advice centres, research centres etc) as well as trade unions transforming their organisation and ideology. All the above arguments suggest that there would be serious obstacles to a more centralised and comprehensive approach.[20]

However even a non-comprehensive and decentralised model would face severe obstacles. A first consideration is that it is necessary to be as clear as possible on the overall economic consequences, or the <u>net effect</u>, of the planning alternative. Since the approach we have been discussing involves a wide range of mechanisms and institutions, this may be more difficult than is immediately obvious, if for no other reason than the fact that policies have multiple effects. Policies geared to a particular goal may have consequences in other spheres. As

Geroski (1988, p22) has argued:

> Instruments generally cannot be narrowly targeted at specific goals because they often have many unintended (and usually unanticipated) effects. For example, R & D joint ventrues (or, in the limit, mergers) initiated by otherwise competing firms may be essential to the development of technology in certain sectors, but may also facilitate collusion over prices... The consequence is that any technology policy designed to solve problems associated with scale economies in R & D may create a monopoly problem as an unintended side effect.

Thus in the above example industrial policy and competition policy cannot be considered in isolation. One can easily imagine other areas of policy where this would also be the case. If one adds to this the fact that policies have differing effects (once again perhaps unanticipated) on various regions, or social groups, then one can begin to envisage the scale of the problem. Thus although the type of planning we have been discussing here is decentralised and non-comprehensive, the question of coordination remains central to the coherence of any alternative economic strategy.

The role of a planning agency, or the DTI in other models, in co-ordination and improving information flows, will not be an easy one. The whole model would be predicated on a much improved flow of information between government departments themselves, and between government departments and enterprises, for both the formulation of a national plan and its execution. We have pointed to the problems confronted by governments in seeking to integrate public spending and nationalised industries within the plan. Furthermore, if one of the goals of planning is to open up economic decision-making to more democratic accountability, one should not underestimate the degree to which 'much of the information required to implement the social audit type of approach ... is not available, and many of the relationships between policy and outcome will be highly uncertain and/or controversial (Gilhespy et al, 1986, p154). Similar problems will be associated with integrating public purchasing power as an instrument in 'commercial leverage' (section 3.3).

The feedback on national planning from decentralised planning experiments also points to certain problems in the co-ordination of the planning alternative. In chapter 1 we discussed the importance of examining the institutional framework through which markets and plans work. Such a framework determines, to a large extent, the manner in which firms compete on a whole range of factors including marketing abilities, product engineering, quality and service. This implies that for planners to have a coherent response to sectoral strategies for development, then their knowledge of such issues must be comprehensive. It means, as Hodgson (1986) has put it, using the knowledge and experience of planning, industrial organisation and financial structures in a way appropriate for economic and social objectives.

If much experience has been gained at the local level, one should not underestimate the problems this will entail if it is transferred to the national level. It will require the type of expertise and size of bureaucracy which will in itself create new problems for efficiency and accountability. Such questions as the control of state holding companies and NPEs, or the internal structure of nationalised industries have been considerably undertheorised in the planning literature. Realistic and workable strategies here, and at the level of state administration, are likely to be an important test for any implementation of an alternative planning exercise.[21]

Two final comments on the overall coherence of the planning mechanisms/institutions are in order here. The first issue is the opposition, which any alternative economic strategy will face, from forces with very different social and economic goals. Only a model of society based on the essential harmony of economic relations can ignore this. The success of the interventionist model will then ultimately depend not only on the coherence of its planning mechanisms but also the extent to which there is support for its project in society.

The second issue relates to the integration of short-term economic policy into the planning framework. We have seen throughout the first four sections of this chapter that this may be of paramount importance for the overall implementation of an alternative economic strategy. In chapter 3.1 we noted that the national plan encompasses both short and long-term economic issues and entails a clear statement of social and economic priorities. For instance a desire to increase the level and quality of investment will have important consequences for macroeconomic policy. It was argued that the balance between consumption and investment is a political question. As Nove (1983, p211) has pointed out while at the micro level profitability may be a good proxy for efficiency, this is not necessarily true at the macro level: 'it would be no evidence of inefficiency if it were decided, by increasing wages, to reduce the level of profits and savings. Of course, if an increase in wages is accompanied by a simultaneous rise in investment and other state expenditures, then the total increase in demand could outrun supply, and the result would be inflation, balance of payments crises, shortages and other unpleasant phenomena.' Thus macroeconomic control is an essential element in the coherence of any alternative strategy. The fact that in the past many alternative economic strategies have been abandoned because of such macroeconomic problems as inflation, or balance of payments crises, further underlines the importance of macroeconomic policy. It is to this issue that we now turn.

While monetarist theories have tended to portray inflation as a basically monetary phenomenon, the political economy tradition has emphasised that it reflects conflict over real resources (see Rubery and Wilkinson, 1986, p328; Goldthorpe and Hirsch, 1978). But as Griffith-Jones (1983) has argued such conflicts either between different classes or groups, or between the public and private sectors, are likely to be intensified after the election of a left government. Such a government usually faces a pent-up demand for improvements in wages and welfare benefits, or the reduction in unemployment, which it is ideologically committed to support. Furthermore, the existence of a tight policy previous to the election of a left government does not entail that inflationary pressures have been removed from the system.[22] On the other hand the problems of the supply-side are likely to be no less acute. Apart from the general uncertainty for investment, it is clear from our previous analysis that the new planning institutions cannot be expected to show an immediate supply-side response.

The existence of such claims on resources for a left government reemphasises the need for economic priorities to be clearly articulated - indeed in chapter one we argued that a harmonisation between short and long-term interests was central to any socialist strategy. The way macroeconomic policy is integrated with the planning alternative will provide a test case for the ability of a left government to promote such harmonisation. In discussing the Chilean case, Griffith-Jones (1983, p7) is critical of the lack of vigour shown by the Popular Unity government in controlling wages. The important conclusion drawn is that:

> the political leadership should make more explicit choices on the distribution of resources between different social groups and economic sectors, given the country's existing constraints... The projection of financial disequilibria - both internal and external - could show the political leadership the need to make urgent choices. If such choices are not made in time, the country begins to live 'above its means'. The resulting high levels of inflation and decline in foreign exchange reserves would threaten any government attempting the already difficult task of pre-transition, in the face of both internal and external opposition.

Thus if the central goal is to raise the level of investment, entailing a favourable climate for such investment, this may require that consumption acts as a 'shock absorber' to ensure internal and external balance (see Hare, 1985, p141). This may mean that improvements in wages and welfare expenditures are introduced only gradually. Correspondingly if this is not politically feasible, then the potential for investment, and the supply-side response expected, should be scaled down.

If such coordination of short and long-term economic policies is not achieved, the resulting disequilibria will have serious consequences, especially if they are perceived to be unsustainable. For instance the fear that they will eventually lead to deflationary policies, with falling demand and higher interest rates, will affect investment, and such planning instruments as planning agreements. After all one of Holland's starting points was that given the 1960s 'stop-go' cycles many long-term projects were not undertaken because firms feared ending up with excess capacity. The case for planning agreements rests on the benefit to be gained from firms adopting more conditional precommitments than they would do otherwise (see section 3.2.3). But this implies that both sides are able to plan over the medium term and thus the government must be able to control its own macroeconomic strategy.

While the planning package we have been examining does not entail a 'dash for growth' through traditional Keynesian reflation, this is not to say that demand factors do not matter.

Our previous analysis has indicated ways in which demand policies could complement those of the supply-side. For instance, we noted the need for an appropriate macroeconomic context if the work of the LEBs was to be one of genuine job creation, rather than job substitution (see section 3.2.2). In this light some have argued for a 'structural Keynesianism' package, where demand-management policies are combined with policies to influence sectors where market failures, or organisational problems constitute an obstacle to any demand-led growth strategy (Thirlwall, 1986; Pesaran, 1986).[23]

The conclusion of the above is that the government must coordinate its macroeconomic and supply-side policies. Even if such a strategy is promoted, its implementation is likely to be far more difficult in practice. For macroeconomic policy will have to respond to external and internal shocks - we saw in chapter 2 how the increasingly unstable world economic context had serious implications for indicative planning in France. Such shocks are particularly significant if they reflect a permanent shift in economic conditions. Hare (1985, p143) notes that adjustment to such shocks can entail both macroeconomic policies to maintain external or internal balance, and some change in the volume and/or composition of investment to adjust on a longer-term basis. The problem is that all too often the macroeconomic response can prevent the longer-term response (eg deflation to restore the balance of payments or raising interest rates). Once again the interrelationship between short and long-term economic policies is of critical importance.

One recent approach aiming to alleviate the conflict between these two sets of policies is that of 'conditional precommitments' which seek to reduce uncertainty: 'although these agents are not assured of a completely fixed policy environment, they do at least know the circumstances under which macroeconomic policy may be changed and how it would be changed' (Hare, 1985, p144). Nuti (1986, p89) has a similar conception in his discussion of 'future contingent instruments': 'ie legally binding unilateral commitments, on the part of the government to adopt at a future given date or dates a given instrument of economic policy (say, a tax of subsidy) and/or a given parameter or parameters for that instrument (or package of instruments) conditional on a given state of the economy (say a given level of employment, or the growth rate of income).' Ellman (1986) mentions a rudimentary form of such an instrument which was used by the last Labour government in announcing certain budget measures subject to a required response by management and labour.

If the conclusion here is that the coherence of the alternative planning model rests on the coordination of short and long-term economic policies, then there is good reason to think that this extends to some form of incomes policy. It is to this that we now turn.

3.6 The political economy of incomes policy

The left has a long tradition of emphasising the importance of planning incomes as part of the wider framework of planning the economy as a whole: 'The common ground between the writings of the academic and political left was a recognition that the problem under discussion was a political one, which required a political solution.... The planning of wages was regarded as just one more element, albeit an important one to be planned along with profits, dividends and other economic variables' (Ormerod, 1981, p184). This follows from the response to economic imbalances, or structural economic problems, in terms of a planning alternative. For such an alternative implies to a certain degree the restriction of the market system, a partial substitution of political mechanisms for market ones. Thus as Crouch (1979b, p20) has pointed out:

> One does not need to accept the Marxist labour theory of value to recognise that ultimately every price reflects the cost of the labour input of the good in question. Any policy which partially suspends market forces needs to find alternative means of restraining prices -ultimately the price of labour... Within the domestic market in an economy with an organised labour movement... control can be secured only by incorporating regulation.

Moreover in the previous section, we noted that economic imbalances, and the conflict over resources between different classes or economic sectors, may be particularly acute on the election of a left government. In this context some form of incomes planning may assist an alternative economic strategy in clarifying important priorities. Any alternative strategy must be anti-inflationary as a whole, at least in the long run: 'to the extent that trade-offs do exist between price stability and other objectives they express an underlying lack of control over some key dimension of the economy' (Grahl, 1981, p210).

However, in recent times, with a wide range of experience of different forms of incomes policies in various post-war economies, the Left has increasingly shown a marked hostility to incomes policies. This has been backed up by marxist academic work which has seen (neo-)corporatism as an essentially capitalist strategy to incorporate workers into the capitalist economic framework (Panitch, 1981). Support for incomes policy is now more often associated with right social democracy, although there has also been some recent theoretical work and proposals on the subject from the Left as well. By examining the different possible rationales for incomes policies we can shed light on the extent to which such policies can be seen as a natural complement to alternative planning proposals.

All economic strategies have, whether implicitly or explicitly, political implications for incomes. It may be thought that this does not apply to monetarist policies which usually reject explicit incomes policies. Broadly, the monetarist view is that the problem of inflation is not endogenous to the economy but a result of government action. Excessive wages cannot in themselves cause inflation, but unions may be important in pushing up wages faster than productivity increases, and then exerting pressure on governments to expand the money supply in the (in monetarist theory, hopeless) attempt to avoid unemployment. The government can thus bring inflation under control by controlling the money supply. However, sociological theorists of inflation have pointed out that such an approach is not very illuminating on the question of why governments allow the inflationary process in the first place (Goldthorpe and Hirsch, 1978).[24] As Goldthorpe (1985) has written the basic weakness of the monetarist case: 'is the assumption that market outcomes can command general acceptance rather than being in themselves a source of dissent and conflict which in turn creates inflationary pressure'. Unless one does assume people will be willing to accept market outcomes, a stabilisation crisis created by monetarist policy to allow market forces to control labour may work, by shifting the balance of power against labour, but it is likely to be a highly temporary solution (Goldthorpe, 1985; Scitovsky, 1978). After almost ten years of the present British government's economic strategy, there still seems little chance to reflate the economy because of socio-economic barriers to reflation (Burkitt, 1982; Bleaney, 1985).

Furthermore the monetarist strategy is necessarily a political one, for with no commitment to full employment it 'implies the underwriting by governments of the outcomes whose acceptability is in question' (Goldthorpe, 1985). In the British case, the strategy has included targeting real wage declines on groups who are in a poor position to organise their own defense, such as the unemployed, part-time workers and employees in small firms (Rubery and Wilkinson, 1986). Thus the monetarist project too entails an implicit incomes policy.

We now turn to examine the extent to which incomes policies are a natural complement to traditional Keynesian economics. Much of the early discussion on incomes policies was linked to the idea developed by Kalecki about the political business cycle and his belief that if full-employment capitalism was to survive it would have to develop new social and political institutions which would incorporate the new bargaining strength of the working class. For Kalecki, if capitalism could not develop in this way, it would leave open the path for socialism (see Kalecki, 1972; Henley and Tsakalotos, 1992; Sawyer, 1985).

In the wage-theorem view (Hicks, 1974; Artis, 1981), incomes policy can be seen as an adjunct to traditional demand management policy and control of inflation by aggregate demand. In the face of labour's ability to maintain the level of wages, a contractionary fiscal policy would have a significant effect via the reduction not in prices but real output. Conversely, when aggregate demand is used in an expansionary manner, the effect is dissipated in a further upturn in the rate of inflation (Goldthorpe, 1984b). As Artis (1981, p18) points out if it is assumed that there is no dispute about real wages and employment, then the rationale for incomes policy can be seen as a 'public goods' argument:

> no individual bargaining group feels (correctly) that its wage bargaining has much influence on prices; each feels that it must match expected inflation plus its desired real wage gain; none has an incentive to wage restraint for this would expose it to real losses. Wage restraint thus has the quality of a public good; if all groups operate within a framework of agreed restraint provided by an incomes policy, then similar real gains can be secured with less inflation.

As Artis also stresses, this rationale takes the employment level as given. If it only took a marginal sacrifice of full-employment to control wages, then this rationale for incomes policy is considerably reduced. Here we are in the world of trade-offs between inflation and unemployment which, via the Phillips curve, seemed to be an attractive framework for economic policy for some time in the post-war period. However, even here incomes policies were portrayed as having some role in either trying to shift the Phillips curve or alter its shape.

As in the monetarist case, this conception of incomes policy is not without its own set of political implications. Shonfield was an early critic of the ability of incomes policies, in coordination with demand management, to solve the problem of inflation once and for all:

> What the fashionable exponents of incomes policies seemed constantly to ignore was that they were asking wage-earners to accept [that] the existing division of wealth and the income derived from it was basically fair... Labour is really being asked to give its consent to a particular status quo... the practical approach to a more rational wage policy must be deliberately and extensively political. (Shonfield, 1965, pp217-19).[25]

Thus in an important sense the introduction of incomes policies is not a natural extension of Keynesian economic policy (Skidelsky, 1979). The social democratic consensus, discussed in chapter one, while it entailed full employment and the expansion of the welfare state also rested on the rights of private capital and a predominantly indirect means of intervention in the economy which did not challenge the prevailing processes of allocation and distribution. The important shift with the introduction of an incomes policy is precisely that the government is in one way or another involved directly in the distribution of income for 'in the context of a money economy, the regulation of money incomes implies the regulation of real incomes; and it follows, then, that incomes policy can never be simply an anti-inflationary device. It must always at the same time represent a political intervention in the distributive process' (Goldthorpe, 1985). In other words, it involves the type of intervention in the economy that it was the object of the previous consensus to avoid.

It is in this context that we can understand the hostility of certain left thinkers to incomes policies which seek to control labour within the capitalist context and disguise the implicit assumptions being made with respect to distributional issues. The experience of incomes policies is that, in the short run at least, they have managed to keep the share of wages in national income at a steady rate which is lower than would have been the case without the policy (see Fallick and Elliot, 1981). For the left this entails that:

> in effect, the government accepts as an institutional datum the saving habits of capitalists and strives to preserve profits; capitalists' consumption remains untouched, whereas the share of workers is adjusted to whatever level is consistent with the equilibrium of aggregate demand. By guaranteeing the profit margins consistent with investment requirements, the state essentially nationalises part of the product, which is handed over to capitalists through a transfer of wealth from one class to another; workers are compelled to restrict their consumption in order to facilitate investment, yet they possess no rights in the ensuing capital accumulation. (Burkitt, 1982, p135).

It is precisely here that the radical implications of incomes policies can be seen, for it suggests that workers will demand a quid pro quo for their cooperation. This process has been described and analysed by Pizzorno (1978) as one of political exchange, where workers offer restraint in the labour market for greater influence in the political sphere. Regini (1984) suggests a number of reasons why unions may want to indulge in political exchange: they may have more power in the political arena than in the labour market (due to weak industrial organisations or a pro-labour party in government); the state may offer more benefits than their employers (social reforms, industrial policies in favour of employment); there may be risks, within the context of a weak economy, in utilising full market power which may further weaken the economy and create greater problems for the unions in the future.

Thus we can agree with Goldthorpe (1985) that there are two rationales for incomes policies. The first seeks to increase government intervention to preserve the status quo with some form of corporatism, presenting the issue as a technocratic exercise in managing the economy. It is a fear that this rationale constitutes the essence of incomes policies which perhaps explains the opposition to incomes policies from some on the left of the political spectrum. The second aims to challenge openly the status quo by accepting and making use of the political implications of incomes policies through a strategy of political exchange. This provides a form of incomes policies that may be accepted by left social democracy, and which significantly, can be integrated into the type of alternative planning we have been discussing in this chapter. The following section discusses what may be entailed in such an approach.

3.7 Social contracts, incomes policies and planning

The radical implications of incomes policies stem from the nature of the quid pro quo that workers may demand for incomes restraint. Traditional theorists, such as Shonfield (1965), seem to imply that political exchange could be limited to such issues as price controls, dividend limitations and policies sustaining growth and employment. However, as Goldthorpe (1985) shows, the logic of such bargaining opens up further questions for:

> as experience of income policies grows, in whatever historical context, those subject to it display a 'learning curve' which leads to greater awareness of the nature and implications of the policy, and then in turn and especially on the part of organised labour, to heightened resistance to it or to a significant enlargement of the conditions for its acceptance.

Thus if incomes restraint will encourage investment, workers are likely to seek some control over the ensuing process of capital accumulation, on the level and type of investment to be undertaken.[26] Moreover, since productivity depends as much on the type of technology used (controlled by capitalists) as much as on effort, workers have a legitimate concern over the type of investment (Burkitt, 1982).

Social contract theories build on such arguments to link control of wages to the wider issues of planning and the socialisation of investment.[27] Purdy (1981) although critical of British experience with the Social Contract in the 1970s, sees that it had many radical implications which were not taken up. He is critical of those on the Left who seem to believe that militancy among workers must always entail pushing for the highest possible wage increases (in the language of 'political exchange' the underutilisation of the market power of labour can be a sign of strength not weakness) and claims that the strategy potentially allowed labour to break the habits and mentalities of corporate defensiveness and widen its social base. For Purdy, two novel features of the British Social Contract were that it widened the influence of trade unions, in particular within various tripartite bodies, to all levels of the economy, and by institutionalising and formalising that influence it developed an ideology of 'no responsibility without power'.[28] Finally Purdy links his discussion to the further issues of economic democracy and planning agreements.

The question of economic democracy focuses attention on whether corporate networks necessarily incorporate workers within the capitalist status quo. As Crouch (1979b, p22) has argued this assumes that:

> a corporatist strategy employed by dominant elites is actually successful. There is one major condition for this success: the organisations which simultaneously represent and discipline the working class have to operate primarily downwards, ordering and controlling their members. If instead they... convey demands to the state and organised capital, not only do they impart a strong element of pluralism, but it is a pluralism which is less constrained by the market and by the institutional segregation of policy and economy characteristic of liberal capitalism.

The strategy of political exchange, then, entails not only the underutilisation of unions' market power, but their ability to develop a politico-economic strategy, and one which has the properties of aggregating their members demands and coordinating them with those of other social groups.[29] It seems that a social contract may well have to rely on a great deal of worker awareness and participation. In chapter one we saw the difficulties involved of any alternative economic strategy harmonising the often contradictory short and long-term demands of workers. It is to such problems that 'political exchange' addresses itself.

The first problem is the need for unions to aggregate their members demands. This may entail, if the state is to have an interest in bargaining, that workers, and especially those with above-average market power, underutilise their power. A good example of this is the Swedish labour movement's strategy of 'solidaristic wages policy' which seeks to close differentials among workers (Martin, 1979; Esping-Anderson, 1984). This will create problems for unions representing workers with significant market power, especially since the exchange for their sacrifice is likely to benefit all workers, as well as other groups such as part-time workers and the unemployed. This demands that workers act as a class, and is linked to a particular labour heritage: 'under the influence of Marxism in its social-democratic variant, these movements saw themselves as acting in the interest of an encompassing working class, and the development of the corporatism of the present day cannot in fact be adequately understood without reference to this emphasis on class as distinct from narrower sectional interests' (Lehmbruch, 1984, p77).[30] It is important not to underestimate the obstacles to such an approach. For as Goldthorpe (1985) points out, for powerful trade unions in a poor economic situation, the alternative strategy of 'dispersing the costs of economic adjustment within a pool of secondary labour, rather than 'internalising' them, is a powerful attraction to union movements even where they possess some tradition of more solidaristic strategies.'

Many theorists have concluded that a high degree of centralisation and concentration of interest organisations is a prerequisite for a political exchange strategy. Regini (1984), however, argues by discussing the Italian experience of solidarieta nazionale, that there can be functional equivalents for representational monopoly. Solidarieta nazionale implied that strong unions of the Italian north would exercise restraint in exchange for both social policy and also the redirection of investment to improve the prospects for women, young people and workers in the Mezzogiorno in general. Thus Regini (1984, p135) concludes:

> on the one hand, the egalitarian and solidaristic roots of their own strategy, as well as the PCI's pursuit of social alliances, led the unions to a set of demands which could appeal to a wide range of social groups... On the other hand, the unions' legitimacy as successful agents of social mobilisation - stemming from the years following the 'hot autumn' of 1969 - was still strong enough to lead most of these groups to delegate to the unions, openly or tacitly, the

right to act on their behalf in bargaining with the government, rather than seeking to act themselves, as independent pressure groups. In this way, then, the Italian unions obtained a <u>de facto</u> oligopolistic position in political bargaining, which helped to make a relatively stable concentration possible.

A second problem with any political exchange strategy is what Pizzorno (1978) has called the interpretation gap. That is, if workers are to offer restraint, they must have some degree of confidence that the package deal will be enforced. This problem is rooted in any strategy which tries to balance the short-run and long-run interests of workers - sacrifices now for benefits later. Some aspects of the package deal can be imposed relatively swiftly, for instance industrial relations legislation. However this is not the case with employment creation or investment strategies.

Here we can see the link between our discussion in this section and the arguments for planning developed in earlier sections for:

> in a mixed economy the framework within which governments regulate policy depends upon a number of variables that they do not control, such as the volume and direction of investment, corporate strategies and the balance of foreign trade. Since failure to determine these variables ultimately undermines the acceptability of prices and incomes policies, this dilemma can only be resolved by a major extension of collective planning of economic activity. (Burkitt, 1982, p137).

In the Italian case, noted above, part of the benefit expected was to include measures on industrial restructuring and reconversion, not only to foster employment but also for increased control over the economy, and redistributive goals. This entailed the government:

> to elaborate 'sectoral plans' to set the criteria for granting public funds to firms who wished to restructure. It required firms which sought to take advantage of public funds to inform the government of their 'programmes of investment and entrepreneurial activity'. And it required public inspectors to certify that assisted firms had preserved previous levels of employment before they could obtain

further funds. All of these provisions were seen as elements of a more dirigiste industrial policy and of planning in accordance with union objectives. (Regini, 1984, p140).

In the event, as Regini observes, many of these benefits did not occur either because of the opposition of the bureaucracy or because the government was not willing to pursue the logic of the policy. Thus the unions' eventual disillusionment in this case has less to do with the ability of unions to aggregate their members interests and integrate them with those of other groups, and more with their estimation of the declining benefits to be expected from such a strategy. The trade-off was not acceptable.[31]

Another approach with respect to the interpretation gap problem can be seen in the Swedish social democrat proposals for wage-earners funds (see Martin, 1979). Here the goal was to link incomes policies not to the type of planning we have discussed but to a form of socialisation of investment. The Meidner committee, which led to the well-known Meidner Plan, saw wage-earner funds as complementing the solidaristic wages policy of Swedish unions.[32] With a solidaristic wages policy there is always the difficulty of persuading high wage-earners, in the face of excess profits, not to seek to push for higher wages. However a system of collective profit sharing could help convince higher wage earners. Such a strategy could also be linked to progressive industrial relations legislation and lead to a deconcentration of wealth. Here the radical implications of incomes policies can be seen, as the quid pro quo for income restraint becomes a strategy which allocates a certain percentage of the profits in the form of newly issues shares to such funds, administered by the unions themselves.[33]

In fact, due to the opposition of the private sector and the lack of the social democratic party's active support, this strategy was never fully implemented. Instead a watered-down version was introduced, which was essentially a stock investment scheme - taxes on profits above a certain level were used to buy stocks. However the potential of this form of political exchange is clear. The neoclassical-Keynesian synthesis has focused on the importance of wages, or wage rigidity, as the central problem to which demand management could offer some solution. However non-orthodox economics has tended to give more attention to the instability of the investment function in a capitalist economy. As Higgins and Nixon (1983), developing the work of

Halevi, have argued:

> inflation depends much more on the direction of capital investment and composition of output than on a change in the relation of wages to productivity. In itself, a high profit investment economy is prone to instability and speculation. In these conditions there exists no necessary relationship between wage and productivity increase... For Halevi, post-war full employment was the cue for 'political trade unionism' to intervene directly in the allocation of resources.

The final problem associated with the political exchange strategy concerns the question of economic power. The question is the degree to which private control of the means of production constitutes an obstacle to the radical implications of incomes policies. This also relates to some of the traditional problems associated with incomes policies, such as wage drift, fringe benefits, and tax avoidance (Fallick and Elliot, 1981). As Burkitt (1982, pp136-7) points out these entail: 'differential advantages in a mixed economy [which] are embedded in the structure of property ownership.' Burkitt summarises well the dilemma for incomes policies:

> they must treat all classes on conventional criteria of fairness if they are to gain acceptance, but if equally fair to wages and profits they defeat their main purpose. Permanent controls on profits lead to the erosion of the private sector and the eventual socialisation of investment, as the government is increasingly compelled to provide output and employment that private capital is unwilling to supply. No effective planning is possible so long as the 'commanding heights' of the economy are in private ownership and under private control... without institutional reforms income control is either spurious (a wage freeze through public relations) or it destroys capitalist motivations without providing workable substitutes.

It is for this reason that there has been such a hostile reaction to, for instance, wage-earner funds in Sweden. Pontusson (1984) does not doubt the radical implications of such measures but questions those over-optimistic accounts which ignore the organised opposition to such developments. In particular he suggests the necessity for such schemes to be promoted at the ideological level to create the type of popular support that they will need. In the British case Leys (1985) has argued that British manufacturing support was converted to monetarism when they grasped the potential of attempts to link elements of economic democracy of the Bullock Report to the Social Contract. Delors' (1978) discussion of French planning points to both the potential of social contracts and the opposition they are likely to face (see chapter 2.3). Two points are of interest here. Firstly, as Delors points out, the failure to reach agreement on incomes virtually destroyed the institution of French planning. And secondly, that unions do have a choice - they can refuse to cooperate or cooperate under such terms which may open up fields of struggle that can challenge some of the mechanisms of the capitalist economic system. Such considerations point to the fact that successful social contracts are predicated on both active popular support and a shift from private to public economic power. The extent and nature of this shift will depend on the particular conception of the socialist transition and characteristics specific to individual countries.

To conclude, we have been examining the interrelationships between the planning of incomes and the planning of the economy as a whole. While we have outlined various proposals under the general heading of political exchange or social contract, we have not looked at the precise details of the mechanisms for controlling incomes. It is difficult to say anything at the general level for a lot will depend on the history and nature of industrial relations, as well as the type of economic imbalances and necessary policies in particular countries. Furthermore, the precise details may not be so important. As a number of theorists have argued, if there is no consent for incomes policies, then most strategies will fail, at least in the long run with the production of ever more ingenious methods entailing an attempt to disguise the true function of incomes policies (see Nuti, 1986, p86); and if the consent is forthcoming then most strategies will have a good chance of working.

We have also pointed to two rationales for incomes policies - that of providing new political mechanisms to prop up capitalism and that which seeks to use them as part of a package deal to challenge capitalism in some form. As Goldthorpe (1985) has written, the key to the first rationale is to limit the radical implications of incomes policies, to restrict money wages, and to incorporate trade unions, and especially their leadership, into tripartite negotiations. The second strategy rests on precisely making the most of the radical implications. But this entails, we have argued, a further condition. For if economic problems, such as inflation, are rooted in distributional conflict, then this conflict must in some way be also an obstacle to any strategy and:

> the argument must be that incomes policy - in much the same way in fact as a policy of monetary strictness - can be effective, if at all, only through bringing dissent and conflict out in the open and their making it evident that any decisive solution to the economic problem will have to be a socio-political one, rather than simply a technical one, and one out of which clear winners and losers will emerge. (Goldthorpe, 1985).

3.8 Conclusions

We have discussed the rationale, scope and coherence of a planning alternative. We need not repeat the arguments here. Rather we present, in schematic form, a planning schema which incorporates a range of institutions and mechanisms which have figured in the literature and which have typically been supported by those in the left social democratic tradition. This cannot possibly provide a blue-print applicable to all conditions and countries, and while we have seen the package nature of these proposals there is clearly scope for varying the appropriate mix of measures, and particular measures may well be more relevant in some specific circumstances than in others. Nevertheless such a planning schema may act as a useful benchmark for our discussion of the Greek case.

A. National Plan

1. Such a plan should facilitate debate within society on alternative economic and social options. Institutions and procedures are needed to promote participation in plan formulation and the consensus necessary for plan implementation. This may require that the costs and benefits of any economic restructuring are seen to be equitably distributed.
2. The plan will outline the central economic and social priorities for the medium-term, providing a framework

for coordinating economic decision-making and clarifying the role of state intervention. It should seek neither to replace the market mechanism nor to be too comprehensive.
3. A central concern is investment - both increasing the volume and improving its structure. This entails: formulating investment plans, by sector and/or region; delineating possible inconsistencies and supply-side bottlenecks; ensuring financial resources for investments which existing firms do not intend to undertake; and facilitating organisational changes for the formulation of new firms, or mergers between existing ones.
4. The coherence of the plan will also rely on such measures as ensuring the administrative and organisational prerequisites for planning, facilitating the necessary information flows within the various levels of planning and so on.

B. Planning Mechanisms and Institutions

The goal here is to go beyond indicative planning to ensure the implementation of the plan objectives, by providing a framework for the incorporation, to varying degrees, of the private, public and social sectors and by promoting industrial policy and strategic planning at various levels. We have suggested a number of issues pertaining to the nature and scope of various planning instruments or mechanisms: the range of such regulatory or interventionist mechanisms needed; the variety of possible ownership forms; the rationale for social ownership and/or intervention and the areas in which it is relevant; the level at which intervention is desirable and effective; and the extent to which these mechanisms can be accompanied by social goals and labour participation.

B1. National Enterprise Boards and New Public Enterprises

Some form of state holding company can contribute to more active planning by: setting up NPEs and joint ventures with the private sector; investing in new technologies, in high-risk areas or where existing firms do not intend to fill the gaps indicated by the national plan; promoting strategic planning at the sectoral level by assisting organisational restructuring and/or acquiring a foothold in a sector for reasons of technology, acquisition of information etc.

With respect to NPEs it is important to consider the overall needs of the sector in which they will operate, their internal organisational structure and their relationship to the planners.

Local planning is important in providing information about the local economy, delineating market opportunities for new investments and promoting an improved developmental environment in terms of R and D, manpower planning, marketing etc. In this context Local and Regional Enterprise Boards can support and/or reorganise local enterprises, revamp managers and management structures, promote social ownership by extending share capital, and identify possible investments and/or create new firms to carry them out.

B2. Planning Agreements, Financial Incentives and Commercial Leverage

Planning agreements have a function in gathering information from important firms/sectors, providing a basis for channelling selective government assistance to industry and increasing the accountability of firms' behaviour. Commercial leverage entails coordinating the multiple instruments that the state has at its disposal in providing leverage over firms: planning agreements, public procurements, extending social ownership and so on. Furthermore there is a need for a careful monitoring of firms' response to the operation of commercial leverage. Packages of support can be provided for sectors or firms at the national or local level including access to finance and markets, provision of management consultancy, manpower planning advice and so on.

C. Complementary Institutional Framework

As well as individual planning proposals we have highlighted a range of complementary measures which improve the overall developmental framework.

A NIB can provide high-risk funding for long-term investments, act as an intermediary between public financing and the private sector, and prepare, in cooperation with the NEB or LEBs, packages of support for sectoral planning. We noted the need for such financing (and complementary services) to be linked to the goals of the national plan and the necessity for transparency in any subsidies for social goals.

There is also a scope for intervening in financial and trade flows, operating some form of exchange controls and providing a framework for controlling MNCs, perhaps at the super-national level. Competition policy should be geared to the needs of investment and organisational restructuring and should be flexible enough to encourage the inter-firm collaboration that planners may wish to promote. Finally we have suggested the need to formulate policies for the increasingly important retailing sectors.

D. Macroeconomic Policy and Structural Change

The national plan will make important claims on available resources. In this context it is crucial for macroeconomic policy to be geared to the medium-term concerns of the plan and to protect planning from external shocks to the economy. We have also suggested that the national plan should have a certain degree of flexibility, avoiding macroeconomic point forecasts and be presented in the form of a rolling plan.

Incomes policies can contribute to macroeconomic control and provide an explicit framework for resolving conflicting claims on resources. Incomes policies can be tied more directly to planning by providing a form of social contract which goes beyond price/dividend controls to encompass the socialisation of investment, social control of financial flows and so on.

The importance of harmonising short and long-term aspects of policy, as well as conflicting claims between social groups and classes, is a crucial component of macroeconomic control and thus of the coherence of the alternative economic strategy as a whole.

Notes

1. As Desai et al also point out the need for such social control over investment is also related to macroeconomic stability:

   > under capitalism, such problems are dealt with at the level of the whole economy in an extremely indirect, veiled manner: it is class conflict which ultimately determines society's distribution of resources between 'consumption' and 'investment', and the outcome of this conflict is then realised through the mediation of the financial sector. Fluctuations in economic activity, often of catastrophic proportions, are the common manifestations of this process.

2. Hare quite rightly points to the tendency of many economic forecasts, for instance in the British Treasury model, to treat investment as a component of aggregate demand at the expense of seeing its role as also relevant to productive potential: 'given the importance of investment for the generation of growth, both in income and employment, this feature makes such models relatively unsuitable for planning processes' (Hare, 1985, p223).

3. It is important to point out that this is to use the term strategic planning in a manner which is distinct from its traditional usage, where it was more often related to macroeconomic imbalance.

4. Cowling (1987, pp13-14) argues that such protection may be defended on infant-industry type arguments. Best points to the extent that Japanese planning has been based on dynamic comparative advantage, whereby international competitiveness can be promoted in the long run by investing in sectors with potential which may not be evident by existing market indicators with existing products, processes or firm organisation.

5. It thus also turns, left social democrats would argue, on the question of economic power (Gilhespy et al, 1986, p81).

6. This statement comes from the British Communist Party's 'Facing up to the Future', a discussion document seen as a preliminary to the redrafting of the party's 'British Road to Socialism' programme (see Marxism Today, September 1988).

7. A more extensive discussion is provided in chapter 6.3 when we discuss PASOK's attempt to 'socialise' public sector enterprises.

8. Much of the debate eventually centred around the development of a particular set of ideas which came to be known as the Alternative Economic Strategy (AES). The AES was concerned with a far wider range of issues than we can hope to deal with here, including the reflation of the British economy, import controls and industrial democracy. Here we are concentrating on the planning proposals. For an extensive debate on the AES, from differing perspectives, see Ward, 1981; Cripps, 1981; CEPG Annual Reports; Glyn and Harrison, 1980; Sharples, 1981; LWG/LCC, 1980.

9. We refer to the model presented in Holland (1975). All references to Holland's work are from Holland (1975) unless otherwise indicated.

10. For the British case it was unfortunate that the lack of implementation of the measures prevented much knowledge being gained from the necessary trial and error process which would have accompanied the introduction of the new planning mechanisms (Coates, 1980).

11. (Holland envisaged existing nationalised industries to be linked to the Cabinet planning committee through their traditional sponsoring department).

12. Many supporting the planning agreements approach have thought it almost a prerequisite that it should be accompanied by an extension of collective bargaining to company strategy, as well as at the level of the central planning institutions (see TUC/Labour Party Liaison Committee, 1982; section 3.2.4 of this chapter).

13. An initial problem concerns the ambiguous nature of the legal status of such agreements, an ambiguity which can be exploited by private sector firms (Hare, 1985, pp237-9). This problem will be more acute the more hostile the private sector is to such planning mechanisms.

14. See Cohen (1969) for a widespread practice of this in French planning.

15. It can be noted that in the case of Greece, there existed in 1981 a National Investment Bank, a highly regulated banking system and an extensive system of capital controls. Given this fact, this section is limited to giving only the broadest outline of some of the necessary complementary policies.

16. An alternative to public ownership for socialising financial flows is the wage-earner funds system debated in Sweden (Meidner, 1978; Martin, 1979). This issue is discussed in later sections of this chapter and also in the concluding chapter of the book.

17. Cooper (1987) discusses the constraints of capital flight in France during Mitterand's first government, especially in the period 1981-83.

18. For a discussion and an empirical study on the issue of capital flight and the extent to which it can be a problem when governments introduce policies which are perceived as hostile to the interests of asset holders, see Gibson and Tsakalotos, 1990.

19. It should be noted here that such conditionality can also be used to ensure other social aims - the idea of contract compliance has been canvassed to ensure that firms with public contracts make progress on equal employment opportunities for men and women (see Gilhespy et al, 1986, pp140-1).

20. See Glyn (1985) for an example of a more centralised and comprehensive approach.

21. As Gilhespy et al (1986, p162) have concluded: 'Competent transparent administration is probably the most important and most difficult part of empowering people'.

22. Such suppressed inflation is likely to be exacerbated by any fall in the currency after the election of a left government (Rubery and Wilkinson, 1986, p334).

23. Structural Keynesianism, and strategies which seek a mix between supply and demand policies, are discussed in chapter 5.

24. Sociologists have pointed to the increasing distributional conflict as a root cause of the current inflation based on such factors as the decline of traditional legitimacy of inequality and the rise in the power of organised labour. Some monetarist do suggest that a root cause of the crisis may be the effect of democracy on the economic system (see Brittan's contribution to Goldthorpe and Hirsch, 1978; Hodgson, 1984).

25. Perhaps this optimism was, as Shonfield (1965) himself suggests, based on the belief that continuous growth could reduce conflict over the division of this new output. Even in this case, the view of those economists, such as Hayek, that concern over relative shares in a growing economy was irrational has been severely undermined by Hirsch's (1977) work on the role of the 'positional' economy.

26. It is unlikely that controls can be limited to dividends and not retained profits, since capitalists will still gain from workers restraint from the increase in the value of their shares (Goldthorpe, 1985).

27. A wide range of approaches is compatible with the social contract idea. For a discussion of the British Left, see Kitching (1983) on preemptive unionism, Hodgson (1984) who emphasises the role of economic democracy in a social contract, and also Purdy (1983), Burkitt (1982) and Hirst (1982). For a critical response see Panitch (1986).

28. As Thompson (1984, p115) points out, the mechanisms of the Social Contract: '...opened up a 'field' of representational institutions that were to be the site of fairly diverse and at times contradictory political and economic struggles within the state itself.' As we argued in chapter 1, the willingness to see the possibilities of 'struggles within the state itself' is a hallmark of social democratic thinking.

29. It has been suggested that one reason for the failure of the British Social Contract was that income restraint by the trade unions was so delayed that eventual agreement of wages was seen more as a crisis measure, after which things would return to normal practices, rather than part of a new socio-economic package (Purdy, 1981; Higgins and Apple, 1983).

30. Lehmbruch (1984, p77) contrasts this heritage to that of revolutionary syndicalism (especially strong in France) with a distrust of tying closely the functions of unions and party, and the experience of British trade unions: 'in which the identification of distinctively class interests was never strong enough to lead to the organisational concentration that is found in Sweden, Norway or Austria.'

31. This aspect may be more important than the problem of relatives, at least within the context of an alternative economic strategy (see Hodgson, 1988, p272). A number of analysts have argued against the inclusion of explicit distributional content in incomes policies, on the grounds that there is unlikely to be widespread agreement on the question of relativities and what there is will be already within the existing structure of relative pay (see Hare, 1985; Fallick and Elliot, 1981; Brown, 1986). However we have argued that any incomes policy has distributional and political implications.

32. The Meidner committee stressed the necessity for a collective approach for individual claims would encourage wage drift and deprive unions of any influence over the funds.

33. Stephens (1979) estimated that in 50-80 years, under such a strategy, a huge majority of Swedish capital would be collectively owned.

4 PASOK's economic alternative

In this chapter we examine the nature and coherence of PASOK's alternative economic strategy. In particular we demonstrate that this strategy, as it developed up to the 1981 election, shares, both in its conceptualisation of the scope of economic policy and in the set of economic policies promoted, many of the characteristics of the left social democratic approach which we have analysed in previous chapters[1].

PASOK (Panhellenic Socialist Movement) was formed in 1974 after the fall of the military dictatorship (1967-1974), and it constituted a novel political formation within Greek politics. This, and the fact that it rapidly achieved power in only seven years, creates serious theoretical problems in delineating the nature of PASOK as a political party. From the start there was those who doubted that PASOK was a genuine socialist party (see Elephantis, 1981)[2]. However we shall leave such questions aside for the moment, returning to them after examining the theory and practice of PASOK.

Here we will be looking at the development of PASOK's economic alternative as it was revealed in various official pamphlets, election programmes and theoretical discussion papers. Particular attention will be given to the contributions of

PASOK's leader, Andreas Papandreou, not only because of his reputation as an academic economist, but also because of his considerable domination over the party apparatus (Koulouglou, 1986; Featherstone, 1987). By examining PASOK's concern with planning, socialisation and general interventionist policies we shall draw on the analysis of previous chapters. In section 4.1 we examine, very briefly, some relevant aspects of PASOK's political strategy. In section 4.2 we go on to examine the central themes of its alternative economic strategy.

4.1 PASOK's political strategy

PASOK's three main aspirations of national independence, popular sovereignty and social liberation, which were considered as interdependent, were first stated in its 1974 'Third of September Declaration' and reiterated in the 1981 election manifesto (PASOK, 1974; PASOK, 1981, p13).

National independence implies: '...a Greece, where decisions will be taken by the people themselves, without foreign dependence, influence and interventions...' (PASOK, 1981, p13). In PASOK's early period this rested on the adoption of the 'centre-periphery' theoretical schema of the dependency school[3]. It coincides with the period of most hostility towards NATO and the EC[4]. While we can see a process of moderation over these issues after 1974, the idea that Greece is a 'dependent' nation is never fully rejected.

Popular sovereignty entailed for PASOK not only the acceptance of a socialist transformation through 'peaceful and democratic' means (Papandreou, 1981, p407) but also a strategy of extending democratic structures to institutions such as trade unions and local government, and within the economy itself. This approach, which we have seen as characteristic of left social democracy, is the context in which to understand PASOK's support for democratic planning and socialisation of the economy which, as we shall see, are both central to PASOK's economic alternative.

Social liberation is a more nebulous concept and is associated by PASOK with a great variety of goals including the decommodification of labour, the ending of alienation and exploitation, and a cultural revolution. There is some confusion over whether these aspirations are means to a socialist end or describe the end point of the socialist transformation.

PASOK expressed its conception of the transition to socialism in terms of a 'third road', explicitly rejecting both traditional social democracy and Leninism, while at the same time accepting what it termed 'non-dogmatic' Marxism (Papandreou, 1977, p37). The rejection of the neutrality of the state was central to PASOK's critique of social democracy[5] (Papandreou, 1977, p37). But PASOK accepted the relative autonomy of the state, that a socialist movement could exploit the contradictions within the state. For this reason: 'the change in the structure of the polity, the state, is an inviolable condition for the radical change of its social structure' (Papandreou, 1978, p17). It will be noted that such a conception also opposes Leninist strategies of dual power discussed in chapter one (Papadatos, 1984, p14; PASOK, 1983, p29-30).

The crucial aspects of reforming the state in the transitional period entailed decentralisation, self-management and socialisation: 'the radical transformation of the state is brought about by the broadening and deepening of the institutions of representative democracy, together with the simultaneous development of popular participatory processes with the creation of forms of direct democracy based on the support of self-managing decentralised processes' (Papadatos, 1984, p15). Our later analysis of PASOK's strategy for democratic planning and socialisation must be seen in this context - both as desirable ends, but also as means by which participatory democracy facilitates a transformation of economic power within society.

Thus PASOK itself considered its emphasis on structural changes as the distinguishing characteristic, the 'touch-stone', of its socialism (Papandreou, 1978, p69). Central to PASOK's conception of structural changes are institutional changes. Here institutions are broadly defined as those mechanisms and functions which constitute the forms of organisation of society and in which are situated relational antagonisms of that society (PASOK, 1983a, p24-5). The main institutions in which PASOK intended to intervene were those of the administrative and coercive apparatuses, the institutions of representative democracy and the new institutions of PASOK for more direct and participatory democracy (agricultural and urban co-operatives, socialised enterprises, neighbourhood councils, democratic planning etc).

This, then, is how PASOK proposed to avoid what it saw as

the danger of 'reformism'. Firstly, it sought to link the policy of a socialist government with the final goals of socialism. Social democracy was criticised by PASOK for limiting its strategy to marginal reforms within capitalism, reforms which are infinitely reversible since they are not protected by structural changes (Papandreou, 1978, p26; PASOK, 1981, p24)[6]. Secondly, it sought to ensure an appropriate rate of introduction of structural reforms:

> A very slow rate involves the danger of the degeneration of the endeavour, the absorption of the structural changes in the framework of capitalist society or their undermining by the untouched strategic centres controlled by the establishment. A too high rate, however, involves the danger of overtaking the institutional possibilities and may lead to the disorganisation of the economy. (Papandreou, 1978b, p70; see also PASOK, 1981, pp24-5).

This stress on the importance of structural interventions also relates to PASOK's conceptualisation of the nature of the economic crisis facing Greece: 'The crisis is structural. The impasse of the Right cannot be avoided, since in order to take the economy out of the crisis, it would have to challenge those structures which are also the structures of its power, to strike at those interests which it has promised to defend and serve' (PASOK, 1981, p57). Thus PASOK did not promote its alternative economic strategy as one merely to modernise the economy. On the contrary, its economic policy was presented as entailing a transformation of economic power. In terms of our earlier analysis it was a project envisaged in a political economy context.

PASOK's strategy rested on a new 'block' of forces, on the alliance of, amongst others, workers, agricultural small-holders, small businessmen and various middle strata of society. This strategy, once more influenced by the centre-periphery schema, was first presented at PASOK's central committee in 1978 (PASOK, 1982a). Greece's peripheral capitalist status entailed that capitalist relations of production were not dominant and co-existed with pre-capitalist forms. Furthermore, the dynamic capitalist sectors were dominated by foreign capital and constituted 'enclaves' with few positive linkages with the rest of the economy[7].

The transition to socialism could not therefore rely exclusively on the working class - they were neither a particularly large section of the population, nor given Greece's dependent social formation, were they likely to grow much in number in the future (Papandreou, 1981a, p12). PASOK's strategy would have to rest on a broader category, of all those who felt shackled by Greece's dependent condition.

How was such a disparate group of classes and interests to be bound together into a coherent strategy? PASOK felt that two factors could facilitate such an alliance. The first relied on the centre-periphery schema: 'based on the common needs and common interests of the exploited social classes, there has developed in our country a new Social Block of forces for change with tangential class interests. The central direction of struggle has been established by the dominant contradiction with imperialism and national dependency' (Papadatos, 1984, p58). In many PASOK accounts this often came to a simplistic analysis that apart from the foreign element and a relatively small group of the national oligarchy which benefited from this dependency, most people in Greece formed part of the alliance of 'non-privileged Greeks' who had an objective interest in allaghi (change)[8]. Secondly, the alliance was to be held together by PASOK's political and ideological role: 'the work of our Movement is to further, within the framework of popular struggles, those political and ideological fermentations which lead to the consciousness of socialism' (Papandreou, 1978, p26; see also Papadatos, 1984, p55, p61). Thus although PASOK clearly dissociated itself from a Leninist approach, it still partly relied on PASOK playing a vanguard role (see Papadatos, 1984, p16).

Two critical problems can be noted here concerning these two factors facilitating the cohesion of PASOK's proposed alliance. With respects to the former, there is the problem that while after 1981 PASOK's use of the 'centre-periphery' schema became increasingly rare, it still held to the same conception of the alliance of 'non-privileged Greeks'. With respect to PASOK's ideological role in promoting cohesion we should point out that this implied the existence of a strong and autonomous PASOK acting as an independent force from any future PASOK government. These two problems will be re-examined after we examine PASOK's actual practice in government and its ability to promote a cohesive strategy. These problems are merely stated here; for in chapter one we argued that a critical component of any left social democratic strategy was its ability to harmonise the interests between the various classes and social groups, and between the short and long-term interests of the alliance as a whole.

4.2 PASOK's economic strategy

PASOK's emphasis on structural intervention and intermediate goals, of linking reforms to the final goal of socialism, is also evident in its economic strategy: 'the treatment of the present unacceptable situation, the overcoming of the crisis, necessitates a series of responsible measures and careful steps. But this does not mean that we shall separate our action into two phases. In other words to limit ourselves to the management of the crisis and then later to begin structural changes' (PASOK, 1981, p23). As we shall see this implied that short-term economic management would have to be closely coordinated with PASOK's structural measures (see chapter 5.3).

This approach entailed a broadening of the scope of economic policy: 'the crisis in the capitalist system on a world scale, with ever-increasing unemployment and inflation, intensified international competition, the increase in the degree of concentration and the monopolistic structure of many branches, have made the traditional means of economic policy ineffective' (Greek Government Programme, 1981, p28; my emphasis). In 1981 Papandreou clarified what was entailed in this wider conception:

> Our own choices are based on the application of democratic planning, the reform of the framework in which public enterprises function, on suitable incentives towards the private sector, on supporting the small and medium sized industries, on clearing up the matter of heavily-indebted industries and on a nationally gainful policy for foreign investments. (Greek Government Programme, 1981, p28-9).

In his speech to the 10th synod of PASOK's central committee, Papandreou (1983c) drew attention to the limitations of both the private and public sectors to carry out some of the major goals of PASOK's alternative economic strategy. He pointed to the organisational limitations of the private sector, and the short-term and speculative character of Greek capitalists who based their activities on maximizing the available subsidies and other forms of state assistance rather than seeking to modernise and restructure their activities. He also pointed to the organisational limitations of the public sector, its bureaucratic nature and its creation on the basis of clientelistic relationships rather than as an instrument for rational intervention in the development process. Such problems both reflected the structural weaknesses of the development of Greek capitalism and constituted a formidable obstacle to any alternative strategy.

Thus Papandreou (1983c, p36-7) concluded that PASOK could not rely exclusively on the existing private and public sectors, but must intervene with a series of institutional and structural reforms. Eventually, as we shall see, the three most crucial areas for intervention were to be codified in three new laws for the socialisation of the existing public sector, the creation of supervisory councils, and the rehabilitation of ailing firms. Such measures were seen to have a dual rationale. By extending participation and social control they were seen to be a good in themselves: 'gradual reform of the structures of the economy so that basic economic choices are made by the social whole' (Greek Government Programme, 1981, p28). On the other hand by transforming the institutional framework such policies would also further the modernisation and development of the Greek economy. Here we will examine the development of such policies as socialisation and democratic planning, before going on in subsequent chapters to the actual record of PASOK in government.

4.2.1 Socialisation

There is a considerable vagueness in PASOK's conception of socialisation, a term that has undergone various shifts of emphasis since 1974. In particular the relationship between socialisation and property relationships was never fully clarified.

Initially Papandreou (1976, p571-2) had contended that the regulation of property relations was not only a legal problem but one which affected the economic, social and political structure of society. For this reason key sectors of the economy had to be socialised. Indeed Papandreou (1977, pp45-6) argued that PASOK's willingness to transform the 'basic social relations of production' distinguished it from traditional social democracy.

In his influential The Transition to Socialism Papandreou follows a line of argument similar to that of Holland (1975). He begins by pointing to the degree of concentration in the Greek economy: in 1975 the 100 largest industrial firms, a mere 4.6% of all firms, controlled 56% of all sales and 51% of working capital and had 53% of all profits (Papandreou, 1978, p71). He uses this as a criteria for the industrial units to be nationalised in order to control the commanding heights of the economy. His discussion on the appropriate level of compensation to property owners suggests that we are talking of a change in property relations[9].

However socialisation was to be clearly distinguished from nationalisation, and initially at least, was integrally linked to self-management. This was seen as crucial to PASOK's promise to transform the relations of production (see PASOK, 1975, p23). On the other hand PASOK was also critical of the Yugoslavian self-management model with its well-known problems with respect to externalities and employment creation. Thus PASOK argued that if a socialised firm was of importance to the national economy, then the central government should be represented on the firms' board, whereas if the firm was of only local importance, the board would have representatives from the local administration (PASOK, 1975, pp18-25; Papandreou, 1981a, p15). Such a formulation implies the existence of conflicting interests, but PASOK nowhere explicitly stated who was to have the upper hand, apart from general statements over the need for conflict resolution to be decided by democratic and participatory means.   The original conceptualisation of socialisation encapsulated two distinct meanings: the extension of social control and the transformation of the relations of production. What was never clearly articulated was the extent to which the latter constituted a prerequisite for the former.

By 1981, PASOK (1981, p105) was still committed to socialisation as 'the basic lever for the radical restructuring of the new strategy for economic and social development'. The list of sectors to be socialised was extensive: the credit system and insurance companies; energy and public utilities; large freight and transport; large foreign trade firms; large mining firms and shipyards; the steel, cement and fertiliser industries; the pharmaceuticals industry; and any enterprises directly associated with defence. However by 1981 we can detect a tendency to separate discussion of socialisation between the socialisation of existing nationalised firms and the socialisation of firms under private ownership (Papandreou, 1981b; Greek Government Programme, 1981).

With respect to existing public enterprises, PASOK intended to extend their management to the workers and representatives of the central or local administration. This entailed that: '...the standard for their aims, their organisation, their operation and their management will serve the social whole and not large private interests as happens today' (Greek Government Programme, 1981, p39). The socialised public enterprises were to be linked to planning by including instruments of control within the firms to harmonise their corporate plans with that of the national plan. Apart from this, socialised enterprises were to have increased autonomy and transparency without the standard Greek practice of using public enterprises for social policy without a transfer of an equivalent amount from public funds (Greek Government Programme, 1981, pp37-8). Thus PASOK's policy was geared to both extending participatory procedures and social control, as well as increasing the efficiency and harmonisation of socialised enterprises with planning.

With respect to private sector enterprises the issue is more complex. By April 1981 Papandreou was arguing that: 'socialisation takes on for every different situation a different meaning' (quoted in Koulouglou, 1986, p55). The eventual policy was not clarified until after the election. By February 1982 Papandreou was proclaiming that PASOK did not intend to nationalise well-operated profitable industries as was done in France (Axt, 1984, pp189-9). The strategy was further clarified during Papandreou's speech to the ninth session of PASOK's (1982b, p27) central committee. Papandreou declared that because of financial constraints, the extensive socialisation previously envisaged would, for the time being, be limited to the creation of supervisory councils, which would act: 'outside the firm, above the firm, but would encompass the workers, the administrative staff of the firm, the local administration and a representative of the planning organ of the country. In this way their [the firms'] course is encompassed into the national plan'. In this formulation the issue of extending social control was being clearly prioritised over that of transforming relations of production. For the supervisory councils schema entailed neither the transformation of property relations nor, necessarily, would they be able to enforce a transformation of relations of production within the firms.

There are two ways of looking at PASOK's eventual formulation over socialisation. The more cynical view is expressed by Kotzias (1984, p362) who argues that the extreme subjectivity of the term socialisation, if it is restricted to mean what is in the social interest, entails that it can take on any meaning the government chooses at any particular time. And certainly PASOK's decision to abandon the state ownership of profitable firms, distances itself considerably from Holland's (1975) conception. A more balanced conclusion perhaps is made by Catephores (1983, pp55-9). He argues that socialisation for PASOK entailed the creation of instruments of control within certain firms in order to supervise policy and harmonise it with the plan: 'the actual force of which may however increase, depending on the determination of economic planners to dictate a certain line of action to business'. Even the cautious nature of supervisory councils 'can also be seen as a beginning of social control that may grow if economic planning proves successful and the economy regains some strength.' This leaves a final assessment open until we have examined the practice of the PASOK government.

4.2.2 Democratic planning

The importance of planning was clearly stated in PASOK's 1981 election manifesto: 'Democratic planning constitutes the guiding and regulating organ of our economic and social policy. It is the guarantee of the unity of direction of our national goals, of the preservation of the general balance and harmonic development of the economy, the link between Central and Local government, as well as between public and private sectors of the economy' (PASOK, 1981, p100). The national plan was seen as the basic 'lever' for development and one of the key institutional changes to transform the structure of society by extending democracy to the economic sphere (Greek Government Programme, 1981, p29).

In early accounts the scope of the plan, and where it would lie on the indicative-imperative axis, was never fully clarified. PASOK (1975) did make it clear that its conception was different from East European central planning, that it did not intend to have quantitative vertical planning, that the market would continue to predominate in the consumption goods sphere and that within certain limits firms would operate with considerable autonomy. Thus Papandreou (1978, p73-4) wrote that: 'entrepreneurial units (cooperatives, socialised and non-socialised firms) will take their decisions within the general framework of the national development plan. The decision criteria of the socialised firms will be established to a degree by the national planning agency. Otherwise firms - socialised and non-socialised - are free to take their decisions.' Here we have a rather enigmatic formula as to the precise balance between indicative and imperative planning, although Papandreou adds that the planning agency will have at its disposal various mechanisms: financial incentives, fiscal policy, price control and shadow prices.

This formula does not seem to diverge significantly from that of indicative planning. Indeed there was to be increasing references from PASOK economic officials on the role of planning in reducing informational uncertainty, in providing a secure framework for investment decisions of the private sector, and thus on the self-reinforcing aspects of planning (Papandreou, 1982; Katseli, 1986). However there are also significant differences between PASOK's conception and indicative planning. It is for this reason that we are entitled to examine PASOK's strategy as a test case for the type of more interventionist planning discussed in chapter three.

Firstly, the emphasis on the decentralisation and participatory aspects of democratic planning was seen by PASOK as crucial to its strategy of avoiding 'technocratic' solutions. This was seen as a clear distinguishing characteristic from both centralised planning but also the social democratic experience. The formulation of the plan, and the social control of its implementation - although the precise details of the latter were left rather vague - were to be decentralised to the regional and local level. This decentralisation was to be accompanied by new 'organs of popular control at various levels' (see Greek Government Programme, 1981 pp29-30)[10].

A second distinction of PASOK's conception from indicative planning stems from Papandreou's (1971) three-fold typology of planning - 'social management', 'development' and 'revolutionary' planning. Social management planning is associated with any planning which does not intervene in the structures of the economy, but which is concerned with the efficient management of the existing system (this is applied to both capitalist and central planning). The second category refers to planning in developing countries and is associated with the analysis of traditional development economics. The third, which Papandreou supports, is revolutionary precisely because it seeks to change the structure of the socioeconomic system. One aspect of this is the decentralisation and participatory aspects of planning. But Papandreou (1980, p331) also argued that this entailed the creation of a new class of investors. New local authority enterprises or agro-industrial cooperatives could play a crucial role in the regional development aspects of planning: 'this sector can cover activities which private initiative or the state machine are incapable of serving' (Papandreou, 1982). This approach is reminiscent of our discussion in chapter 3.2.2 on local planning and the need for local planning boards. For PASOK such institutional intervention is seen as a critical component of its overall economic alternative.

A final distinction from indicative planning was that PASOK economists stressed the importance of the social control of investment as the major rationale for planning rather than Pigovian arguments over market imperfections and the static efficient allocation of resources (see chapter 4.1; PASOK, 1977, p28). This social control was aimed at confronting the cyclical instability of the economy as well as transforming its structure. Thus as well as the proposed five-year national plans, PASOK argued for a more 'active' form of planning: 'Active planning involves the formulation of a credible investment programme to increase present and future productivity, to diversify the productive base and to mitigate inequities and distortions' (Katseli, 1986, pp68-9). It is to this conception of 'active' planning, which was to accompany democratic planning, that we now turn.

4.2.3 Active planning and industrial strategy

'Our basic aim is to enrol industrial activity into a global plan aimed at balanced regional development, technical modernisation, the support of viable units, the creation of new particularly high-technology units and the harmonious insertion of these units into the natural, urban and cultural environment' (Greek Government Programme, 1981, p45). The need for a vertical integration of industrial production and the creation of complexes of self-complementing industrial units was also emphasised.

PASOK's strategy was based on its structural analysis of the problems of the Greek economy, and in particular in industry, reflected in the 1970s phenomena of a decline in productive investments, the return to traditional and less capital-intensive activities and increasing macroeconomic imbalances (see Papandreou, 1979, 1980; Lazaris, 1981; Vaitsos, 1986)[11].

PASOK's response to these structural problems entailed a considerable increase in public investments (see Papantoniou, 1981). Through the national plan framework, the intention was to promote a large coordinated programme of public investments in certain key areas. Furthermore this in turn would necessitate a reform of the public administration and the socialisation of the existing public sector enterprises since PASOK argued that past governments had used no obvious criteria for project appraisal for new investments relying instead on *ad hoc* methods (see Papandreou, 1980, p31-2).

Apart from this autonomous increase in public investments PASOK's strategy also rested on promoting investments under joint public-private ownership and integrating private sector activities into the national plan (Korliras, 1986, p37). Papandreou (1982) envisaged primarily two mechanisms for harnessing the private sector to the national plan. The first was through the supervisory councils discussed earlier. The second was through institutionalising planning agreements (see chapter 3.2.3). These would act to rationalise the financial aid given to private industry and to help the private sector sort out their own needs and potential. Agreements could be reached on specific production or export programmes, an increase in employment or productivity, or the adoption of new technology. Such agreements, Papandreou (1982) considered, in a formulation strikingly reminiscent to that of Holland (1975), constitute: 'a step further on from indicative planning without becoming coercive'.

Vaitsos' (1986, p83) formulation of PASOK's strategy to reverse Greece's deindustrialisation is closer still to Holland's conception: 'We will do so through a process of active planning, planning which involves direct participation at a mesoeconomic level, whether sectors, or large enterprises. When specific packages are presented, the rules of the game will be present and clear and investment will be undertaken'. Vaitsos further argued that industrial strategy should include a sectoral element, with sectoral plans clarifying the potential for planning agreements. Both Vaitsos and Katseli (1986) promoted the need to rationalise public procurements in order to provide some leverage over private sector firms.

A further test for PASOK's more interventionist stance would be provided by the problem of the heavily indebted 'ailing firms'. These included firms, in nearly all industrial sectors, which faced increasingly severe financial problems during the 1970s[12]. In its 1981 electoral programme, PASOK's strategy had not been fully worked out, but had three main aspects. Firstly, it would undertake studies to examine which firms had a viable future. Secondly, PASOK envisaged the radical restructuring, and where necessary a re-orientation of production, of those firms deemed viable. The financial restructuring of viable firms would entail a full, or partial, capitalisation of their old debt. Finally, there would also be a 'decisive' element of worker participation in the management of these firms (see PASOK, 1981, p60). By 1982 Papandreou was making it clear that what was envisaged was a new state holding company to restructure these firms and act as an arm of PASOK's industrial strategy. While noting the scale of the problem inherited from the previous government, Papandreou suggested that PASOK's response constituted an interesting social and economic experiment (PASOK, 1982b, p27). It is an experiment which we examine at some length, in chapter seven.

The 'ailing industries' phenomenon highlighted for PASOK the problems of the Greek banking system and the state's use of financial incentives. PASOK argued that previous governments had never integrated the financial system into any specific strategy for development (Papandreou, 1979, 1980; PASOK, 1981, p69). Furthermore PASOK argued that this was crucial to understanding the nature of Greek inflation - it was not the increase in credits as such that was inflationary but the increase of credit used for unproductive activities (Lazaris, 1981; PASOK, 1981, pp58-9). As Vaitsos (1986, p82) pointed out '...what in fact has happened is a privatisation in the behaviour of state-controlled banks. What we are concerned in is obviously not the ownership, it is the control and conduct of operations. Both industrial and 'problematic' [ailing] firms, as we call them, express this organic relationship between state-controlled banks and private firms and the privileged access to the credit system...' This privileged access, PASOK felt, operated at the expense of the small and medium-sized firms, which PASOK considered of paramount importance in any future development (PASOK, 1981, p68; Papandreou, 1981b).

PASOK's response rested on the socialisation of the banking system. Here socialisation takes on yet another meaning - social control to ensure that finance should be based on productivity and developmental criteria and that it should be in harmony with the goals of the national plan (see chapter 3.3.1). Special mechanisms for financial incentives were also to be created for the small and medium-sized firms (Papandreou, 1981b).

We have seen here that PASOK's conception of 'active' planning, to accompany democratic planning, shares the left social democratic emphasis on intervention to promote the supply-side response of the economy. PASOK's alternative economic strategy can thus be usefully compared with the planning schema presented in chapter 3.8. It remains to be seen how these supply-side measures were to be integrated into the more short-term economic management aspects of PASOK's policy.

4.2.4 Short-run economic policy and structural change

In PASOK's 1977 manifesto little attention was given to short-run economic policy (PASOK, 1977, p41). After the 1977 election, Papandreou set up the 'Committee for Analysis and Planning' (EAP), under the economist Lazaris. Its duty was to prepare a more detailed government economic programme. The discussion papers of EAP have never been published, and it is clear that the Committee had an uneasy relationship with the party apparatus (Koulouglou, 1986, p54)[13]. In two crucial meetings held before the 1981 elections between Papandreou and his advisors, there were serious disagreements over the macroeconomic stance for the first year of a future PASOK government. Some argued that given the scale of the macroeconomic imbalances and PASOK's commitment to structural reforms, the government should begin with a stabilisation policy with only modest redistributive policies. However the majority, it seems, favoured a more expansionary and redistributory policy.

This is reflected in PASOK's 1981 manifesto. On the one

hand it noted the scale of the macroeconomic imbalances, aggravated by the 'scorched earth' policy of New Democracy which had led to serious public deficits (PASOK, 1981, p58). And although it also emphasised the primary importance of structural reforms, this was considered compatible with a redistributory social policy and expansionary macroeconomic policy (PASOK, 1981, p23, p79). Boosting demand, through strengthening the purchasing power of the popular classes, it was felt, would support the activities of domestic industry. This initial 'push' was to be supported by a number of complementary policies including promoting productive investments, prioritising those which offered a quick return, use of spare capacity and import-substitution to protect the balance of payments, price controls and tax reform (PASOK, 1981, p59). Very little detail was given on the precise workings of these policies.

The above rationale for an expansionary macroeconomic policy owes something to the remnants of the dependency schema in PASOK's thinking. In opposition PASOK had accused the right-wing governments of promoting inflation while, on the one hand, restraining the incomes of the popular classes, that basically consumed domestically-produced products, and on the other failing to control the incomes of other classes (profits, self-employed and in general large incomes) that consumed luxury and imported goods, thereby further weakening domestic production (Papandreou, 1978b, p5). This approach had expansionary implications for any alternative macroeconomic strategy, although elsewhere Papandreou sounded a more cautious note. Thus in a 1980 budget speech, Papandreou, while repeating PASOK's charges of the inflationary consequences of government policy, argued that whereas people were ready for sacrifices as an element in an alternative strategy, they must first be assured that the costs and benefits are equally shared.

However expansionary implications were also in evidence in PASOK's highly oppositionist stance, between 1974 and 1981, where it seemed to suggest that the demands of the various sections of the alliance of 'non-privileged Greeks' would be met immediately on the election of a PASOK government[14]. There exists here a contradiction between PASOK's more theoretical concerns over the transition to socialism and PASOK as a political/electoral organisation. The 1981 manifesto is characterised by a long list of promises with virtually no attempt to specify priorities (Axt, 1984, p196).

And yet a number of influential PASOK economists had argued for a cautious macroeconomic policy in the early years of any future PASOK government. Papantoniou (1981) argued that the basic problem entailed an increase in the share of investment which could not be achieved by an expansionary fiscal and monetary policy which was likely to lead to inflation and problems with the balance of payments. Great care was needed over the control of consumption which necessitated a consensus over wage policy with small initial increases within the context of an economic programme to take Greece out of the economic crisis[15].

This more cautious approach was supported by Arsenis (1981). He pointed to the link between past failures of socialist governments and their initial expansionary policy. He stressed the need for a socialist government to allow for unforeseen events and operate a financial policy which could provide some autonomy vis-a-vis the international lending authorities. A socialist government must not be in too great a hurry to achieve power, and must make clear the need for sacrifices in the transitional period. Such an approach parallels that discussed in chapter 3.5 where we examined the need for an alternative economic strategy to gear its macroeconomic and incomes policies to the supply-side or structural policies[16].

As a party, however, nowhere does PASOK explicitly make clear the need for the planning of incomes, as discussed in chapter 3.6 and 3.7. There were frequent references to promote consensus: '...our message to working people is very clear, we ask for vigilance, patience and self-control. From our side, we guarantee, with the institutions that we are creating, that the results of economic progress will be distributed in a socially just manner, rewarding the labour of working people' (Papandreou, 1982). While such an approach does seem to include elements of what we have termed 'political exchange' this did not entail for PASOK an institutionalisation of debate or explicit negotiation with unions, and others, over the course of its macroeconomic policy.

Thus there existed within PASOK a level of unresolved tension concerning the appropriate macroeconomic strategy. How this was resolved is discussed, at some length, in chapter five.

4.3 Conclusions

We have seen in this chapter that PASOK's economic strategy has strong similarities with the left social democratic model discussed in earlier chapters. By broadening the scope of its economic policy, PASOK acknowledged the political economy implications of its strategy - it entailed a transformation of economic power. The emphasis given to institutional reforms, such as democratic planning and socialisation, should be seen in this context. PASOK also recognised that beyond extending democracy and social control to the economic sphere, its strategy would have to promote economic efficiency and development. To do this it relied on the type of mechanisms and institutions, such as planning agreements and state holding companies, which we have argued are characteristic of the left social democratic approach. Thus PASOK's ability to carry out its alternative economic strategy successfully constitutes an interesting test case of the general approach.

We can conclude here by outlining a number of areas of ambiguity and unresolved tension which will be important to the development of the analysis in subsequent chapters.

PASOK's highly oppositionalist political mode up to the 1981 elections suggests certain populist characteristics. The core of the 'populist' argument is that PASOK brought together a whole series of interests who could unite in opposition, but whose contradictory demands would become obvious when the new government had to implement a particular economic policy (see Elephantis, 1981; Papagiannakis, 1980; Mouzelis, 1980). To take just one example: would PASOK's middle class allies, and in particular the self-employed, accept increased taxation, a necessary expression of solidarity since, as PASOK itself acknowledged, tax evasion and the reliance on indirect taxes constituted one of the most formidable structural problems of the Greek economy? In this respect the ability of PASOK to gear its macroeconomic policy to its supply-side policies would be a fundamental test of the overall coherence of its strategy. Macroeconomic control, we argued in chapter 3.5, is an essential instrument to ensure that economic priorities are clarified and that the longer-term aspects of economic policy are sustainable. What is at stake is PASOK's ability to promote a global strategy, one that harmonises both the interests within its supporters, and between the short and long-run interests of the alliance as a whole.

The existence of such a global strategy would enhance the capacity to overcome the opposition that any alternative economic strategy is likely to face. If PASOK was right to argue that the political Right was incapable of bringing Greece out of its economic stagnation, because to do so would mean damaging those sectors of its own support that gained most from the existing model of development, then it follows that those interests were unlikely to remain indifferent to the promotion of a different model of development. This entailed, among other things, clarifying the role of the private sector and the new 'rules of the game'.

Papandreou made it clear on numerous occasions that such new institutions as planning agreements did not imply a threat to the private sector but a new role in the context of Greece's development. But he also made it clear that if the private sector refused to co-operate and invest, then the state itself would be forced to take up these investments (Kotzias, 1984, p428; Korliras, 1986). Arsenis (1981) also stressed the importance of a new socialist government clarifying its intentions early on so that the international community and the domestic private sector were made aware that they faced a determined government and adjusted their activities accordingly. The likely response of those opposed to the new strategy is likely to be a wait-and-see policy, testing the resilience of the new government. If government 'threats', such as its intention to carry out investments that the private sector refused, were seen as a bluff, the private sector was unlikely to treat seriously other aspects of the policy such as planning agreements.

Not surprisingly PASOK tended to scale down the potential for conflict in its strategy as the elections of 1981 approached (Elephantis, 1981b, p205). Increasingly the emphasis was on promoting the broadest possible consensus (Greek Government Programme, 1981, p6). However consensus is one thing and the opposition of those with no interest in the new strategy is quite another. How PASOK responded to such opposition would be another test once it began to implement its strategy.

PASOK's strategy encompassed both state-led initiatives and more decentralised elements of planning 'from below'. This shows some sensitivity to the problems of 'statism' and control which we discussed in chapter 3.2.2 in the context of the increasing emphasis on decentralisation of left social democratic models of planning. Thus PASOK included in its approach both 'active' state planning, reminiscent of Holland's approach, and decentralised planning and the new role for local authority enterprises and cooperatives.

Both aspects were likely to confront formidable obstacles barely mentioned in PASOK's account. State planning was likely to confront the notorious bureaucracy of the Greek public administration (Tsoukalas, 1986). Although PASOK continually stressed the need to confront this problem, little detail was given on how it was to be tackled in practice. To take a couple of examples: given the lack of any serious administrative and research infrastructure, how was PASOK going to enhance the quality of state investments, as opposed to a quantitative increase, and indeed, which areas were the most profitable for investment? The lack of concrete proposals on these issues was likely to take up valuable time once PASOK was elected. Furthermore would planners desire or have the capacity to enforce certain lines of action either through the various organs of planning or within public and socialised firms?

The more decentralised aspects in PASOK's planning proposals were likely to confront the severe weakness of civil society in Greece (Mouzelis, 1980; Tsoukalas, 1986). The dominance of the state in the Greek social formation throughout the twentieth century has resulted in very weak horizontal organisational linkages: trade unions, local councils and pressure groups have been vertically integrated into the state apparatus, and have found it difficult to act as autonomous independent pressure groups. This obviously has serious implications for any strategy of implementing participation in planning, local planning and cooperatives. Would for instance, trade unions accept new institutional measures as a process in which their role was transformed from a defense of economic interests to that of intervening in the sphere of production? Would local authorities find a new level of autonomy in order to develop their own development strategies, thereby reducing their reliance on state dictate?

Thus to conclude, while we have seen that PASOK by 1981 did have a strategy which encompasses many of the themes of left social democracy, there remained a number of areas of ambiguity and unresolved tension. The success of its strategy would, in part, rely on the resolution of such problems and in part on the relevance of the left social democratic model for the Greek case.

Notes

1. Thus there are certain limits to the scope of our analysis, given the concerns of earlier chapters. We do not examine the whole domain of PASOK's economic policy - little, for instance, is said on agricultural policy.

2. See also Andrianopoulos et al (1980) for a wide variety of views on the nature of PASOK.

3. The centre-periphery schema was most important in PASOK's early phase and is not dealt with here in any detail (see Papandreou, 1978; Hadjigregoriou, 1979). For a more sophisticated treatment of the dependency issue, which closely follows the work of de Palma (1978) and which signifies PASOK's changing emphasis, see Zachariades (1980).

4. On PASOK's early uncompromising stance against NATO and the EC see PASOK (1977) and Papandreou (1978). By 1981 PASOK had shifted its position on the EC in favour of a referendum (which was never carried out) and seeking a special agreement (see Greek Government Programme, 1981).

5. Social democracy was initially used by PASOK in its pejorative sense, which we have attempted to avoid. PASOK's use is more akin to our use of right social democracy.

6. This approach can be usefully compared with our analysis in chapter 1.1. In its more theoretical literature PASOK conceptualised this strategy as one of intermediate goals (see Papamichail, 1978, 1982, 1983; PASOK, 1983, pp311-12).

7. For a more sophisticated version of this approach see Mouzelis (1978).

8. For the earliest, and still most devastating critique, of PASOK's conception of 'non-privileged Greeks' see Elephantis (1981). Elephantis develops this critique to examine the populist nature of PASOK.

9. The most radical statement of socialisation in terms of transforming property relationships can be seen in PASOK, 1977, p29.

10. In more theoretical accounts this participation was linked to a strategy of transforming the structure of economic power:

> the central ingredient of a development plan, the key point, can be found by whose interests the plan serves and who controls its formulation and implementation. Planning at the service of the people means above all the substantial intervention and efficient control from organs of popular power in both the planning and implementation of the development strategy. (PASOK, 1983a, p45).

    For democratic planning as an instrument promoting socialist consciousness, see Papamichail (1978; 1983).

11. A more extensive account of the nature of these economic imbalances and structural problems will be given in subsequent chapters (in particular, see chapter 5.1, 6.3, and 7.1).

12. The cause of the 'ailing industries' phenomenon are discussed in chapter 7.1.

13. Thus our account here is based on published articles of PASOK economists (especially at the 1980 Panteios conference on the 'Transition to Socialism') and on various personal interviews with those responsible for PASOK's economic policy.

14. For a critique of PASOK's populism, see Papagiannakis (1980), Elephantis (1981) and Koulouglou (1986).

15. Papantoniou (1981) also favoured import controls as part of any alternative economic strategy. He rejected that this would lead to inefficiency as it would be part of a programme for restructuring Greek industry (see also Lazaris, 1981, p333). Arsenis, in a paper given at the same conference in which Papantoniou expressed this

view, doubted the efficacy of this proposal, suggesting that it would lead to illegal contravention and black markets, as well as jeopardising PASOK's alliance with the middle classes. Furthermore such an approach may entail ignoring the vital issue of tax reform, which he considered a more just and efficient method of dealing with imports (Arsenis, 1981, p79, pp81-2).

16. Papamichail (1983) argued that PASOK must take into account that its strategy entailed a heightened level of 'class struggle' and that therefore there existed limits to the distributional options open to a socialist government focusing on institutional/structural reforms.

5 Macroeconomic policy and PASOK's alternative economic strategy

In this chapter we examine PASOK's macroeconomic policy up to the June 1985 election. Given the emphasis so far on supply-side policies it may seem strange to begin the analysis of the Greek case with macroeconomic policy. This is justified on two counts. Firstly, an examination of macroeconomic policy will indicate the nature of the economic imbalances faced by the new government and provide the general economic context for the interventionist policies to be discussed subsequently. Secondly, PASOK, at least after the summer of 1981, articulated a conception of integrating macroeconomic policy with the other aspects of its alternative economic strategy. In chapter 3.5 we argued that a critical component of any such strategy was the integration of short-term economic management with the planning alternative. Macroeconomic control was essential in order to clarify the economic and social priorities of the government and to preserve the long-term sustainability of the strategy. Thus PASOK's strategy of 'stabilisation through development' and 'gradual adjustment', which we examine here, is a good test case of a left social democratic government's ability to provide such macroeconomic control.

In section 5.1 we set the scene by outlining the

macroeconomic imbalances faced by PASOK in 1981. In section 5.2 we examine PASOK's initial economic policy, which was reversed in the summer of 1982. Then in sections 5.3 and 5.4 we provide a critical analysis of the rationale and implementation of the strategy of 'stabilisation through development' and 'gradual adjustment'. We reach some provisional conclusions in section 5.5.

5.1 Macroeconomic imbalances in 1981

PASOK was well aware in October 1981 that it would face a severe economic crisis. In March 1982 the government submitted a memorandum to the EEC (arguing for 'special arrangements' in the light of Greece's severe economic difficulties) clearly outlining the scale of the macroeconomic imbalances: the rate of inflation had increased in 1980 and 1981 to around 25%, double the rate of previous years and the average of the Community; the rate of increase of GNP had been on a declining trend and the figure was negative for the first time in 1981; the current account deficit had doubled and in 1981 rose to 6.5% of GNP despite the continuing recession; and the public sector deficit increased dramatically to 17% of GNP in 1981 (see Greek Memorandum 1982, p.91).

Table 5.1
Selected Macroeconomic Aggregates

	1975	1976	1977	1978	1979	1980	1981
Growth of GDP[*] (%)	5.1	5.5	2.9	6.4	3.6	1.9	-0.2
consumer price index	13.4	13.3	12.2	12.6	19.0	24.9	24.5
Current account deficit (%GDP)	4.6	4.1	4.2	3.0	4.9	5.6	6.5

[*] at factor cost and constant prices
Source: Ministry of National Economy, OECD

The most serious deterioration can be seen in the period 1979-81 (Table 5.1). In part, this reflects the second oil crisis of 1979 and the effect of Greece's entry into the EEC. However, the performance of the Greek economy in this period was significantly worse than that of other western economies. This is particularly the case for inflation and the public deficit (Table 5.2). PASOK argued that these developments reflected the underlying structural weaknesses, inequalities and imbalances within the economy (Greek Memorandum, 1982, pp90-1).

Table 5.2
General Government Financial Deficit (as % GDP)[*]

	1977	1978	1979	1980	1981
Greece	-6.2	-6.3	-5.3	-5.9	-12.8
EEC	-3.3	-4.0	-3.6	-3.5	-4.8

[*] Reliable figures for the Greek net PSBR for the period 1977-81 are not available.

Source: OECD, Ministry for the National Economy

Underlying the figures of Table 5.1 is a deterioration in the structure and pattern of growth. From 1974 onwards services contributed an increasing amount to the growth figures.[1] Deleau's study on industrial investment in Greece points to the deteriorating investment picture after 1974: 'from a base level of 100 in 69-70, it reached a peak of 150 in 73-4 but declined afterwards to level out at 120 in the '80s' (Deleau, 1987, p5). The share of efficiency-raising investment in GDP (in machinery, equipment and construction excluding buildings) fell from 13% in the early 1970s to 9% at the end of the decade while the tradable sector of the economy declined (OECD, 1983, p9). As Vaitsos had pointed out the 1960s era of Greek industrialisation was being partially reversed in the 1970s. Whereas the former period was associated with an improving sectoral balance with more than one third of investments in intermediate and capital producing sectors, by the 1970s 'Greece was slipping into the traditional sectors' (Vaitsos, 1986, p78).[2] Furthermore, as Deleau stresses in his comparative study of Greece and other industrialised countries, Greek industrial specialisation: 'relies, much more than in most western economies, on products with slowly growing markets' (Deleau, 1987, p8). The overall picture of investment in manufacturing can be seen in Table 5.3.

These trends are in part due to the lack of any coherent development or industrial policies in the 1970s (see OECD,

1982, p9-10; Dedousopoulos, 1981). A more detailed analysis of the structural problems faced by the Greek economy will have to await our discussion in subsequent chapters where we shall examine the appropriateness of PASOK's policy response in various areas (industrial development, restructuring of the public sector etc). For the moment however we can state that the Greek economy was in a weak position to respond to the worsening world economic context of the 1970s and the rise of the Newly Industrialising Countries (NICs) who compete in markets in which Greece has traditionally specialised.

Table 5.3
Private Investment in Manufacturing
(constant 1970 prices)

	Total Investment (Drachma bn)	Yearly Increase (%)	% of GDP
1974	14.8	2.8	4.6
1975	13.0	-12.2	3.8
1976	12.8	-1.5	3.6
1977	12.4	-3.1	3.3
1978	11.1	-10.5	2.8
1979	13.0	17.1	3.2
1980	14.2	9.2	3.4
1981	13.2	-7.0	3.2

Source: Ministry of the National Economy, OECD

The oil price increases by shifting relative prices threatened whole sectors of Greek industry, fuelled inflationary pressures and widened the current account deficit. Furthermore the deterioration of the world economy and the recessionary policies of most OECD countries had a number of further indirect effects. For Greece's previous economic development had relied quite heavily on a 'soft' balance of payments constraint due to the availability of large foreign exchange revenues in the form of shipping, tourism and remittances from Greek workers abroad. Spraos (1984) has argued that the specificity of Greece lies exactly in the existence of this plentiful foreign exchange revenue which, by allowing a high exchange rate, has made Greece a high wage country compared to the NICs but not a low enough wage country to compete effectively with the advanced capitalist countries. However, tourist receipts slowed down after 1979 as a result of the OECD recession, as did remittances (see OECD 1982, p28).[3] Demand management policies since 1975[4] were overall expansionary and there was little adjustment to the 'oil price shock' of 1979. While the New Democracy government attempted to tighten fiscal policy in 1979, they were not successful. Monetary policy was also more expansionary than planned after 1979 (OECD, 1982, p31-40).[5] This policy was associated not only with rising public sector deficits but also a worsening structure of public expenditure. Central government fixed investment fell as a share of GDP from 3.25% to 2.25% between 1975 and 1981.

Thus there were three crucial aspects to the deteriorating economic performance of the Greek economy: the failure of the private sector to invest in industry, the lack of a development strategy since 1974 and the nonadjustment to the worsening world economic conditions of the 1970s. Papantoniou's view, coming as it does from a leading PASOK economist, sets out succinctly how PASOK conceptualised the failure of economic policy before 1981:

> the adjustment of the Greek economy to the energy crises dictated an economic policy with the key targets of reducing the consumption deficits of the public sector, the increase of productive investments, the modernisation of industrial structures and the weakening of inflationary pressures. However, in the first years after 1974 the exact opposite occurred: the deficits expanded, productive investments remained stagnant, Greek industry instead of being modernised became over-indebted with the result of a large number of 'problematic' firms, while inflation reached very high levels. (Papantoniou, 1986).

5.2 PASOK's initial economic policy

Here we examine PASOK's initial economic policy up to the summer of 1982. The overall effect of its macroeconomic policy was expansionary, although this is the net effect of an expansionary incomes policy and a more tight fiscal-monetary policy. By the summer of 1982 the expected results had not materialised and there was a partial reversal of policy.

The redistributive incomes policy is the hallmark of PASOK's first year of government. It introduced guidelines in an attempt to reverse the decline in real wages between 1979-81 and to boost demand (see Table 5.4).[6] The basic idea of the new policy was that to reduce inequality wages would be increased (on January 1, 1982) in inverse proportion to their level, according to the following scale[7]:

workers with contractual monthly earning (excluding family allowances)	Monthly raise (known as 'corrective amount')
up to 20,000drs	5,000drs
25,000drs	4,500drs
30,000drs	4,000drs
35,000drs	3,500drs
40,000drs	3,000drs
45,000drs	2,500drs
50,000drs	2,000drs
above 52,000drs	no rise

PASOK also introduced a partial indexation scheme, known as ATA from its Greek initials, which entailed increases at four-month intervals linked ex-post to the consumer price index. Thus on May 1 there was full indexation for the portion of contractual monthly earnings up to 35,000drs; 50% indexation for the portion between 35,001drs - 55,000drs; 25% indexation for the portion between 55,001drs - 80,000drs; and no indexation above 80,001drs.[8]

The OECD estimates that the net effect of the above was that earnings in the non-agricultural sector rose by 27% in 1982, that is 5.5% in real terms, while, given the scaled nature of both the corrective amounts and indexation, wage differentials decreased. Furthermore the increases in the manufacturing sector reached 37.5% in 1982, with a particularly significant effect on profits and competitiveness of import-competing and export sectors (see OECD, 1983, p17-18). Kapsis has estimated, taking into account taxation increases and inflation, an increase of 3% in total disposable income for 1982 (Kapsis, 1983; see Table 5.4).

It was hoped that this boost to demand via income redistribution would support domestic industry and thus lead, in the first instance, to the revival of the economy (see chapter

4.2.2; Greek Memorandum, 1982, p91). This revival was also to be supported by the promotion of productive investments, at first in those sectors whose returns could be quickly materialised, partly through a new incentives law (Law 1262) and expanded credits to the small-medium sized industries. Irrespective of the specificity of the Greek context it should also be pointed out that this expansionary approach constitutes a 'classical' beginning for a socialist government.

Table 5.4
Average Earnings and Disposable Income
in Real Terms, 1979-86
(Total Economy excluding agriculture)

	Average earnings before taxes		Average disposable income of wage and salary earners	
	Rate of change(%)	1974=100	Rate of change(%)	1974=100
1979	-0.2	134.4	-1.8	130.9
1980	-3.4	129.8	-1.2	129.3
1981	-0.4	129.3	-2.5	126.1
1982	4.2	134.7	2.8	129.6
1983	-2.5	131.3	-3.7	124.8
1984	3.9	136.4	3.1	128.7
1985	1.3	138.2	2.0	131.3
1986	-7.0	128.5	-8.1	120.7

Source: Sambethai (1986), based on Bank of Greece figures.

A more cautious approach can be detected in the budget policy for 1982. The Finance Minister Dretakis, pointing to the economic imbalances, presented his budget as a transitionary one, a first attempt to put state finances and expenditure in order, before the expected introduction of the five-year plan in 1983 (see Commercial Bank of Greece, 1982). The budget in an attempt to reduce inflationary pressures aimed at reducing the PSBR by dramatically increasing revenues (by 58.8%) and by a smaller increase in expenditure (35%) than in previous years. However, certain social expenditures were to expand in line with the government's redistributive social philosophy. In the event, this policy was relatively successful with public expenditure not overshooting and the public sector borrowing requirement being reduced. However tax revenues fell well short of the target, a phenomenon which was to be a recurring feature of PASOK's term of government. The shortfall was mainly due to a considerably slower growth rate than the 2.5% originally forecast and failure to make any significant progress on the much heralded campaign on tax evasion.

The overall policy stance however revealed itself in the poor economic results of the first 6 months. Aggravated by the recent entry to the EEC, the increase in demand did not lead to a recovery of growth, and particularly serious was the increase in import penetration and poor export performance (OECD, 1983). There was a marked increase in the consumption of imported consumer goods (Bank of Greece, 1983, p11). Even with the more moderate fiscal and monetary policy of 1982, the effect of the incomes policy added to the lack of serious readjustment in the Greek economy since 1979 implied a poor supply-side response, strong underlying inflationary pressures, declining competitiveness and a fall in private investment (OECD, 1983, p7-10).[9]

Here we can observe the problems associated with a strategy which attempts to increase production by increasing wages. Greece is a small open economy with structural problems and a wide variety of market failures. Arguments suggesting that a redistribution of income will increase domestic production (because the weaker economic classes consume domestically produced goods) are probably of little validity for Greece.

The result of all this was a cabinet reshuffle in the summer of 1982 which brought G Arsenis to head the Ministry of National Coordination, now renamed Ministry of National Economy (MNE). Papandreou in an interview in 1983 made clear that a shift of position had taken place on the appropriateness of Keynesian type arguments for the Greek economy. Here he claims that the redistribution of 1982 was not part of an economic strategy but part of a policy of social justice, of showing the 'socialist colours' of the new government and that economic policy was a preparation for the introduction of the five-year plan in 1983 in which PASOK's overall strategy would be made clear (see Papandreou, 1983b).

However, before going on to examine the rationale of the new policy, two points should be made. Firstly that, as we shall see, the redistributive elements of the initial policy were to have a long-term effect on the Greek economy. And secondly since it was seen that PASOK's initial policies led to the change in direction in the summer of 1982, PASOK, and not New Democracy, would have to shoulder the political costs of the new policy (Arsenis, 1987, p70).

5.3 'Stabilisation through development and gradual adjustment'

5.3.1 The rationale of 'stabilisation through development'

As is often the case when dealing with a particular government's economic policy, there is considerable confusion between the stated policy of the governing party, the pronouncements of those responsible for economic policy and the actual policy carried out. Here we examine the intended policy of PASOK between 1982 and 1985, as it is revealed in various policy documents and theoretical papers, before going on to look at the policy actually implemented.

If there was a coherent approach to macroeconomic policy in PASOK's first term, then this was one which came to be called 'stabilisation through development' (Arsenis, 1987, p91). PASOK fought the 1985 election on this approach and the expectation was that it would be continued in PASOK's second term (see Athanasopoulos, 1985). It was not until the announcement of the October 1985 stabilisation measures that this policy was brought into question.

The underlying theory of this approach explicitly challenged monetarist and deflationary policies, or at least their appropriateness to Greece. Arsenis (1985a) argued that macroeconomic imbalances of excess demand could not, and should not, be countered by a policy aiming at reducing demand and redistributing income towards profits. For Arsenis argued that such a traditional stabilisation is presumed to operate on an object (economy) which is by nature in equilibrium and has only cyclical deviations. However Greece was, in PASOK's view, in a long-term crisis and therefore in need of policies aimed at changing the structure of the economy and promoting development (see also Papandreou, 1983a). Furthermore deflationary policies cannot easily be imposed by a socialist government committed to improving income distribution, alleviating unemployment and introducing better social services

(Korliras, 1986, p36).

Korliras employed a traditional fix-price model (Malinvaud, 1977) to analyse the Greek economy. His conception was that Greece faced a 'classical unemployment' equilibrium, characterised by high wages with respect to full capacity utilisation, quickly adjusting inflationary expectations and pessimism concerning future prospects. He pointed out that a restrictive demand-management and incomes policy may do no more than shift Greece into an 'underconsumption' or 'Keynesian unemployment' situation with lack of demand and poor growth prospects. He concluded that:

> To avoid such an eventuality a cautious or non-permissive demand and incomes policy must be combined with an aggressive public investment program and .... supply-orientated measures. (Korliras, 1986, p38).

Katseli (1985a) also stressed the need to put most emphasis on medium to long-term supply side policies. She argued that devaluation as a policy response was of limited value since by increasing domestic costs it could lead to a devaluation-inflation spiral especially in Greece where 80% of imported goods are either intermediary or capital goods (limited substitutability between domestic and foreign capital goods), and where there is an unbalanced opening of the economy and a concentration of import structure and currency revenue. Thus the structural problem of the balance of payments needed long-term development policies. Devaluation could provide a breathing space but could not by itself guarantee the needed increase in investment and restructuring to ensure an improvement in the structure of exports and currency earnings (Katseli, 1985a).[10]

Thus, as Arsenis (1985a) often expressed it, 'stabilisation cannot be considered separately from development'. Given the structural nature of Greece's problems, long-run stabilisation would depend on promoting supply-side policies and planned intervention at the industrial level. Crucial to all this was raising investment and Arsenis argued that in the first instance this would have to be led by public sector investment, to be followed later by the private sector (Arsenis, 1984).

But this supply-side policy was to be explicitly integrated with demand management policy. As Korliras (1986, p37) expressed it:

> Although the reduction of inflation remains a policy target,

it cannot be pursued on the mere expectation that market forces will eventually react to bring about the recovery of the economy. On the other hand, there is no room for the traditional expansionary macroeconomic policies, because of the constraints imposed by the high inflation and the deficits in the budget and the external balance ... the government is compelled to adopt a complicated policy package. Rejecting the short-term trade-off between inflation and unemployment; it must strive to reduce both by a mixture of demand-management and supply-oriented policies.

This formulation clearly rejected any simple Keynesian approach. On the macroeconomic side the emphasis was on 'gradual adjustment' (Papandreou, 1983a). Katseli (1985b) argued that such gradualism was the 'cornerstone' of PASOK policy, for shock treatment would run the risk of destabilising expectations and aggravating economic imbalances whereas if:

> a medium-term programme is instead credible, the government can stabilise expectations through consistent action and eventually enhance further its own credibility.

Thus macroeconomic policy should be geared to supporting the medium-term programme of the government expressed in the five-year plan.[11] Prudent demand management policies should contribute to a gradual reduction of inflation, while fiscal policy should aim at restructuring government expenditure away from current expenditure and subsidies towards public investment. The goal was that macroeconomic policy should reduce the current account deficit, as a percentage of GDP, to 3.5% over the medium term (see Katseli, 1985). As Arsenis (1985a) pointed out the policy of reducing external and internal imbalances was linked to the need to find the adequate resources for development. Foreign borrowing could only work as a complementary form of finance if the creditworthiness of Greece was not to be put at risk (see Commercial Bank of Greece, 1983, p115).

5.3.2 The underlying economic theory of 'stabilisation through development'

Before going on to examine the economic policy actually carried out by PASOK after 1982, a small digression is needed to examine the economic theory underlying the strategy of 'stabilisation through development'. For while this strategy could be said to represent a consensus view amongst PASOK economists at the time, it is also clear that the participants in the debate came to their stand-point from various theoretical traditions varying from the orthodox economic toolbox of Korliras to the 'ECLA' or 'dependency' theoretical approach which underlies much of Arsenis' thinking.

Furthermore in their analysis we can detect differences of emphasis, to put it no stronger, which left certain crucial issues open to interpretation: how important were 'high' real wages in the decline of Greek investment since 1974 and, thus correspondingly, how much emphasis should be placed on a non-permissive incomes policy, at least in the short-run? How important in the demand-management side of the strategy was reducing the net PSBR as opposed to merely restructuring public expenditure away from consumption and towards investment expenditure? The lack of clarity, as we shall see, was to have significant consequences for the eventual implementation of the policy. Here we can provide an analytical framework which will facilitate an understanding of the nature of these problems.

Firstly, the question of wages and incomes policy. Korliras, we have seen, by referring to the 'classical unemployment' nature of the Greek economy clearly placed more emphasis on wages than Vaitsos (1986) for instance. The question arises that if in Greece wages were too high: too high with respect to what?

One way of looking at this would be to put emphasis on real unit labour costs. Paulopoulos' (1986) study has shown that in Greece between 1955 and 1974 real wages and productivity increased at about the same rate. However, 'between 1974/5 and 1982/3 the real wage increased by 39.4% compared to only a 9.9% increase in productivity'. He also points out that after 1974 unit labour costs in manufacturing increased faster than the real price of manufacturing products, implying an increase in the real unit labour cost in the manufacturing sector. Table 5.5 tells the same story of worsening competitiveness in marked contrast to the rest of the EEC. Real unit labour costs can be defined as:

$$l = W.L/P.Q$$

where l = real unit labour costs
W = nominal wage rate

L = quantity of labour input
P = price level
Q = quantity of output

A strategy to reduce real unit labour costs, and thus increase competitiveness does not imply a reduction of real wages, merely that labour productivity (Q/L) increases at a faster rate than real wages (W/P). Thus a strategy to improve competitiveness could rely on improving Q/L which depends on, among other factors, investment, technological progress and organisation of production.

Table 5.5
Real Unit Labour Costs (1961-73=100)

	1961-73	1975	1981	1982	1984
Greece	100	90.2	106.4	106.1	107.2
Europe (12)	100	107.2	103.3	103.0	101.1

NB: an increase in the indicator entails a reduction in competitiveness.

Source: Eurostat et Services de la Commission, EEC

The above is a useful analytical framework for discussing PASOK's rejection of a strategy of deflation or reducing real wages, and supporting a coordinated package of demand-management and supply-side policies. For PASOK's supply-side strategy was aimed at increasing investment in industry, promoting technological innovation and restructuring the organisation of production. This would, by increasing labour productivity, improve the competitiveness of the economy as long as the real wage did not rise at a faster rate. Indeed PASOK increasingly stressed not only macroeconomic control but the need to improve labour productivity (see Papandreou, 1983c, pp38-9). Furthermore such a strategy was particularly important since the lack of competitiveness of the Greek economy was not only due to price but also non-price factors, such as quality, service, research and development, which would also be influenced by supply-side policies. This approach can be usefully compared to that of strategic planning discussed in chapter 3.1. The less expansionary and more controlled demand-management policy envisaged in the 'stabilization through development' strategy can then be seen as an attempt to provide a stable macroeconomic framework within which the supply-side policies could be given the time to work. Thus wages could rise, even if slowly, and the brunt of adjustment would not have to be taken up by wage-earners.

A similar approach underlies Arsenis' and Katseli's view about the limits of devaluation for a country like Greece with severe structural problems. Consider Corden's (1981) well-known two-sector model (a tradeable and non-tradeable goods sectors) which can elucidate the policy options open to a country facing a balance of payments constraint. A restrictionary fiscal and monetary policy reduces demand for both sectors and thus helps reduce internal and external deficits. However, since overall demand has also been reduced, unemployment could result since there is no incentive for the tradeable sector to expand, unless, in Corden's terminology, a 'switching' policy is also employed. This can take the form of a devaluation or tight incomes policy. This will shift resources to the tradeable goods sector. However, Corden's model can be considered a static one - the supply-side increase of tradeable goods comes from reducing the real wage. In a more dynamic model, the increase in the supply of tradeable goods could come from the type of supply-side policies supported by PASOK with correspondingly less adjustment coming from changes in the real wage.

Pesaran's (1986) conception of 'structural Keynesianism' can help develop the same point. He discusses the limits of a typical package of devaluation and incomes policy as an alternative to monetarism in the British context. As Pesaran points out there is no need to take such typical problems as a high propensity to import or unfavourable trade elasticities as given. Pesaran points to the weakness of Keynesian economic theory to come to grips with specific market failures which may hinder a supply-side response and to assume that once aggregate demand is at an appropriate level, equilibrium will be reached. As Pesaran (1986, p171) concludes:

> a cut in the propensity to import rather than a cut in real wages should form the basis of an alternative economic strategy. Demand management policies alone cannot ensure such an outcome. Supply-side policies for directing resources towards investment and production of import substitutes and exports, must be integrated with the Keynesian demand-management policy paradigm. In short we need more rather than less intervention in the economy if we are to deal with possible market failures effectively.

The above discussion provides some kind of a framework for understanding PASOK's strategy of combining a cautious demand-management policy with their interventionist supply-side policies. It also helps in understanding some of the differences that have been discussed. The appropriate stance of demand management or incomes policy would depend on how quickly one would expect the supply-side policies to begin to work. For instance, the more optimistic one was on the speed of the supply-side response, the less the need for an adjustment in real wages. Or to put it another way, there were differences of opinion over the effect of wage increases on the competitiveness of the Greek economy in the short run, even it there was general agreement that long-term and sustainable competitiveness depended on the more medium or long-term supply-side policies.

Some care is needed concerning this small digression into the economic theory underlying PASOK's approach. The above discussion is not presented as <u>the</u> underlying theory behind PASOK's economic policy, but in order to provide a framework facilitating analysis. We have pointed to differences of emphasis amongst PASOK economists and politicians, reflecting in part the use of various theoretical models and practical experience. As we shall see these differences also reflect different political and social priorities. The degree of cohesion of a party in implementing a policy in a coordinated and consistent manner is not to be assumed, but has to be tested in the concrete case under examination.

5.3.3 The policy of gradual adjustment

We can now turn to how the strategy of 'stabilisation through development' and 'gradual adjustment' was implemented between 1982 and the 1985 election. This new approach can be seen in the policies adopted after the summer of 1982 and especially in 1983.

<u>Incomes Policy</u> Arsenis announced the tighter incomes policy for 1983 in December 1982, pointing to the need to recover some of the loss of competitiveness of the previous year and to tackle the problem of unemployment (quoted in Commercial Bank of Greece, 1983, p115). More attention was now being paid to the effect of wage increases on the cost of production

and the decrease in investment in 1982 (see Tables 5.2 and 5.11).[12] Thus for 1983 the government's incomes policy was to limit wage increases to the ATA (indexation) payments, while the income tranches of the 1982 scheme were not adjusted. Moreover the indexation increments were 'heterochronised' (delayed) - half the indexation award due on January 1, 1983 was paid on May 1, while the May 1 award was paid together with the September 1 award. This policy was made mandatory by law for both public and private sectors (Sambethai, 1985, p9).

This approach was to accompany the policy of depreciation and devaluation of the drachma in an attempt to limit the effects of devaluation being dissipated in higher wages. A policy of depreciating the drachma was followed from August 1982 leading to the official devaluation of 16% in January 1983. However after January 1983 the authorities attempted to peg the drachma to the dollar, evidently not expecting (in line with most observers at the time) the dollar to appreciate. The actual appreciation of the dollar mitigated to some extent the effect of devaluation, although the drachma was unpegged in August (see OECD, 1983, p34).

Papandreou (1983b) made it clear in an interview that PASOK promised to make up the loss of earnings towards the end of the year, but hoped that the breathing space offered would encourage private and public investments.[13] Thus for 1984 and 1985 incomes were to bear relatively little of the weight of 'gradual adjustment'. In 1984 there was a return to the system of guidelines, indexation increases were not deferred and indexation tranches were adjusted upwards to take into account price increases; although the guidelines did not envisage any increases beyond the ATA payments.[14] A similar policy was followed in 1985, although income tranches were not adjusted. Thus in 1985 Papandreou (1985a, p34) claimed that although wage increases beyond the ATA payments would risk closing thousands of small-medium firms working at the margin, and thus increase unemployment, PASOK's incomes policy, together with certain tax reductions, meant that there would be an increase in the real disposable income of workers. As Arsenis (1985a) was to point out after the 1985 election the strategy of 'stabilisation through development' entailed not a reduction in income but a 'socially just' incomes policy, although wage increases were 'planned' to be compatible with the other targets set by PASOK. Implicitly, therefore, it was being argued that such an incomes policy was not an obstacle to the general state of the economy or to the successful implementation of the supply-side strategy.

Fiscal and Monetary Policy Thus the main emphasis of 'gradual adjustment' was intended to fall on monetary and fiscal policy. Central to this was a commitment to restrain monetary expansion and to reduce the net PSBR by a gradual amount year by year, thereby limiting the public sector contribution to inflationary pressures. There are numerous statements from a wide variety of PASOK officials that this remained the strategy for 1983, 1984 and 1985 (see Commercial Bank of Greece, 1983, p19; OECD, 1986, p18). However, there is some ambiguity over this. For it is clear that some placed less emphasis on reducing the overall net PSBR as such and more on merely restructuring public expenditure towards investment as opposed to consumption expenditure. There is not easy resolution of this at the level of policy pronouncement, as there are policy statements (even by the same persons) which would support either of the two interpretations.[15]

What were the mechanisms envisaged to bring down public sector deficits? The first problem is that public expenditure as a percentage of GDP in Greece is not particularly high, if compared to EEC figures. Public consumption expenditure (public administration, defence, health and education) in 1980 constituted 16.4% of GDP in Greece, compared to an EEC average of 18.1%. In 1980 pensions made up 5.7% of GDP in Greece, compared to an average of just over 10% in other EEC countries (see Stournaras, 1987).

Furthermore, any reduction of the deficit from the public expenditure side was made difficult given PASOK's incomes policy and its commitment to social expenditure and increasing public investment. Incomes of the public sector employees make up a large share of total current expenditure and apart from the incomes policy described above there were certain special features of Greek public sector employment that tended to increase the rigidity of the total wage bill, notably the operation of bonuses based on time served in the public sector and the operation of the 'unified wage scale'. Social expenditures in such sectors as education and health, which were not at a high level compared with EEC standards, were also planned to grow, the latter being associated with the creation of a National Health System. Furthermore expenditure on pensions was bound to increase given the 1982 extension of pension coverage and rights. There was also a deterioration in the age structure in this period. As the OECD points out the social security accounts turned into a moderate deficit in 1982 and the prediction (which turned out to be correct) was for this trend to accentuate in future years unless a determined effort was made to reduce social security tax evasion (OECD, 1983, p49).[16]

Given the above, much would depend on efficiency savings in central government expenditure and reducing the deficits of the public corporations and organisations (DEKO, from the Greek initials). The former was bound to be a long-term affair and would ultimately depend on some form of reform of Greece's notoriously inefficient and bureaucratic public sector. The latter was to be achieved through two mechanisms. Firstly, by restructuring and modernising public corporations (see chapter 6.3). Secondly, prices of public utilities would have to be increased in line with costs. Rather than relying on the tradition of manipulating prices as an implicit social subsidy, PASOK aimed, in those circumstances when it thought such a subsidy justified, to make it both explicit and transparent, the cost being payed from the general budget. The price adjustments of public utilities in 1983 was a step in this direction (OECD, 1983, p47-8).

The result of the difficulties discussed in drastically restraining public expenditure is that a serious attempt at reducing public deficits would entail a significant increase in revenues. Greece combines a relatively low average tax burden (tax revenues as a share of GDP) and relatively high tax rates. For instance in 1984 the average tax burden in Greece (tax revenue plus social security contributions as a share of GDP) was 35.23% compared to an EEC average of 41.93% (Stournaras, 1987). This reflects the small tax base. Taxes on incomes and expenditures which are easily taxable (wages, consumer goods) are high whereas there are enormous problems concerning taxation of the self-employed, the black economy and certain groups, notably farmers who pay virtually no tax at all. Added to this we have the phenomenon of tax evasion, which one noted Greek economist has called a national sport (Spraos, 1986; see also Angelopoulos, 1986, p31-8).[17]

PASOK was fully aware of the size of this problem. Combatting tax evasion and increasing the tax base was seen as a vital prerequisite in ensuring that the cost of policy was equally shared and that all citizens felt that they had a share in the economic strategy being carried out. In nearly all Budget debates, policy announcements by economic ministers and PASOK political propaganda, this commitment was reaffirmed (see for instance, Dretakis, in Commercial Bank of Greece, 1982a, p7). It was thus an essential component of the strategy of 'stabilisation through development'.

5.4 An assessment of the gradual adjustment strategy

In this section we turn to the question of the success of the strategy of gradual adjustment as it was implemented in 1983-85. PASOK itself was to argue that macroeconomic policy had not been as successful as it had hoped and that it was this that necessitated the October 1985 stabilisation package. And certainly a first look at the figures in Table 5.6 would suggest that the policy of gradual adjustment was unsuccessful. Inflation and unemployment do not show any great improvement, while the figures for the current account and public deficit show an actual deterioration. However, as is often the case, the interpretation of these figures has been highly contested. It is part of our task here to attempt to assess critically the policy as implemented. We believe that this debate is important and has clear implications for our assessment of PASOK's overall strategy. Is it the case that gradual adjustment was implemented but contingent circumstances and a failure of political will led to its abandonment, in which case we would have to examine why PASOK changed its policy in October 1985. Or is it the case that the policy was reversed because gradual adjustment was not implemented consistently, in which case the focus of our analysis should be on the reasons why gradual adjustment was difficult to implement in practice and what general conclusions we can draw from this. In this section, in fact, we argue that the latter position is closer to the truth. We do this by focusing on public deficits, which are a crucial aspect of gradual adjustment, given the fact that as we have seen it was intended that demand management policy rather than incomes policy should bear the burden of adjustment. We argue therefore that the public deficit targets were not met and then examine why this was the case. We then go on to discuss the implications of this failure and show that it is confirmed by the macroeconomic picture in 1985.

We begin by examining the argument that gradual adjustment was not as unsuccessful as might appear at first glance. Arsenis and Katseli have argued that a careful examination does not prove that the macroeconomic performance between 1982 and 1985 led to a 'crisis' in the summer of 1985 and that, more or less, the same policy should have been continued after the 1985 election.[18] For instance Arsenis (1987, pp223-4) has claimed that by 1985 both growth and inflation were developing as planned.[19] If these were only gradual improvements then this was because that was the intention. What is more this gradual improvement had been achieved within a context of improving income distribution between 1982 and 1985.

Table 5.6
Selected Macroeconomic Aggregates

	1981	1982	1983	1984	1985	1986
Inflation %						
Consumer price index	24.5	21.0	20.3	18.5	19.3	23.0
GDP deflator (at market prices)	19.6	24.2	20.1	20.3	17.8	19.1
Growth of GDP at factor cost and market prices (%)	0.2	0.6	0.4	2.9	3.2	1.4
Current account deficit						
$ billion	2.4	1.9	1.9	2.1	3.3	1.7
% GDP	6.5	4.9	5.4	6.5	9.8	4.3
Net PSBR						
% GDP	14.3	12.5	11.3	15.5	17.7	13.9
Drs Billion	302	328	351	588	810	761
Unemployment						
% of labour force	4.0	5.8	7.4	8.1	8.5	8.7

Source: OECD 1987

Arsenis does accept that the public and current account deficits were a more serious problem. On the public deficits, he writes that the Ministry of the National Economy was aware that the targets for 1985 were not going to be met, and thus a 'corrective' package of public sector (though not investment expenditure) cuts were presented to the Cabinet on July 7 (Arsenis, 1987, p225). On the balance of payments, both Arsenis and Katseli have argued that the worsening situation was mostly due to speculative activities. Thus Arsenis points out that exports after 1985 were doing well but that there was not a corresponding inflow of foreign currency, as many exporters delayed converting their earnings into drachmas, concerned both with the political crisis in March (in which Karamanlis was not re-elected President) and the uncertainty concerning the coming election. The latter also explains the increase in imports, especially after July 1985, as part of a speculative stock accumulation. As Arsenis also points out disposable income and private consumption did not shoot up in 1985. Therefore, the argument is that if the government had confirmed its continuing commitment to the same strategy, with no devaluation, and had adopted the proposed expenditure cuts, eventually foreign earnings would have had to be converted and importers would have had to reduce stocks. Imports would have fallen in the second half of 1985 and the first half of 1986, making the target - for the period 1985-86 overall - realistic. Arsenis is also critical of PASOK's delay in coming to a decision over economic policy, between the July election and the October measures, which further enhanced speculative forces (Arsenis, 1987, p226-9).

In the opinion of Katseli (1985b):

> We should not confuse short-term deviations from particular targets with long-run adjustment and the validity of a programme. These deviations, when properly analysed and understood, should be corrected through appropriate policy instruments and through corrective measures that strengthen rather than weaken medium-term policy.

We can begin our assessment of the validity of the above argument by examining in a little more detail the degree to which the policy of gradual adjustment was consistently carried out. For whatever the interpretation of the 'crisis' in the summer of 1985, it seems evident, that there are significant divergences between intended policy and outcome, especially in

1984 and 1985. For instance for 1984, the PSBR was projected to decline slightly while in fact it increased considerably from 11.3% of GNP in 1983 to 15.5% in 1984. A similar overshooting is in evidence in 1985. Monetary policy as a result was also more expansionary than planned. Thus in 1984 domestic credit expansion was planned to increase by 19.8% and increased by 26.2%, while for 1985 it was planned to increase by 20.5% and increased by 26.5%. Similar overshooting can be detected in M3 aggregates (OECD, 1987, p12).[20] The Governor of the Bank of Greece was particularly scathing of the 1985 result:

> The chief characteristic of the economic developments for 1985 is the important deviations which were observed in the basic aggregates of the Greek economy from the targets that had been set or the forecasts that had been made at the beginning of the year. (Bank of Greece, 1986, p15).[21]

How do we account for this failure to meet monetary and fiscal targets? A first consideration is the continual importance of general government expenditure in accounting for public sector deficits. For instance, the salary component of public employees in central government, banks (mostly state owned) and the DEKO was very difficult to control.[22] PASOK seems to have continued a long-standing tradition of using employment in the public sector in an attempt to mitigate the increase in unemployment:

> After remaining constant for four years, general government consumption increased from 15.75% of GNP in 1980 to 20.5% in 1985. This was largely due to the rapid growth of the wage and salary bill from 11% to 14% of GNP. Government employment continued to expand at over 3.5% per annum, the same rate as recorded between 1975 and 1980. In the ten years to 1985, its cumulative rise was almost 50%, thus probably exceeding the genuine demand for new posts. (OECD, 1987, p40).

Obviously some of the increase in employment was due to the need to cover the government's new commitments in social services, especially health. But it is doubtful whether this can explain such an increase. And together with the incomes policy operating in the public sector (augmented by the 'unified pay scale' and bonuses for time served and certain education qualifications) this phenomenon made it difficult for PASOK either to reduce the aggregate public expenditure or to restructure public expenditure towards investment expenditure, as Table 5.7 shows.

DEKO prices in 1984 and 1985 were once more kept low, which together with the increasing deficits of the social insurance funds, already noted, contributed to rising DEKO deficits.[23] If we add to this certain subsidies, such as to farmers, it is not difficult to see how public expenditure targets overshot. However part of the explanation of failure to meet targets can also be explained by tax revenue shortfalls. Thus two-thirds of the failure to meet the 1985 Government budget deficit target was due to shortfalls in tax revenue, with both indirect and direct taxes being responsible, as well as the absence of an efficient mechanism to fight tax evasion (OECD, 1986, p22). PASOK's commitment to widening the tax base and combatting tax evasion was not translated into concrete policies.[24] Indeed the policy of raising tax allowances, while not irrational in itself, limited the degree to which the tax base was expanded. The self-employed and farmers continued to pay a disproportionately small percentage of total taxes and many tax exemptions continued to operate (see Angelopoulos, 1986). It is indicative that measures for broadening the tax base and combatting tax evasion were still seen as being crucial when the October 1985 stabilisation measures were announced.

Table 5.7
Breakdown of Government Expenditure 1981-85
(in billions of drachma)

	1981	1982	1983	1984	1985
Consumption	588343	721335	879646	1113254	1497277
%	85.40	85.96	83.87	83.84	84.52
Of which personnel expenditure	204042	273577	333207	428466	570329
%	29.77	32.60	31.70	32.11	32.19
Investment	97080	117851	169211	221043	274197
%	14.16	14.04	16.13	16.57	15.48

Source: Figures based on Bank of Greece (1984, 1986, 1987)

We need to dwell a little more on the question of PASOK's policy on public deficits and the implications of not controlling them. Let us take the national income identity:

$$Y = C + I + G + X - M$$

where: Y = National Income
C = Private Sector Consumption
I = Private Sector Investment
G = Government expenditure (both consumption and investment expenditure)
X = Exports
M = Imports

This can be rewritten as:

$$Y - C = G + I + X - M$$

$$\Rightarrow S + T = G + I + X - M \quad \text{since } Y - C = S + T$$

$$\Rightarrow (M - X) = (G - T) + (I - S)$$

where: (M - X) = current account deficit
(G - T) = public sector deficit
(I - S) = investment gap

This shows that if (G - T) is rising, as we have shown to be the case in Greece from 1974 to 1985 (see Tables 5.2 and 5.6), the current account deficit (M - X) must be increasing unless the investment gap (I - S) decreases. The latter entails either a reduction in investment and/or increasing savings.[25] In actual fact savings in Greece show a small deterioration, as a share of national income since 1979 (falling from about 25% to 20% in 1986). As we shall see (Table 5.11) private investment, especially in manufacturing fell during the PASOK years. Although hardly an encouraging development, it could be argued that this was not necessarily too serious since PASOK's strategy rested, initially at least, on an expansion of public investment, which it was hoped would, at some time in the future, pull up private investment as well. However while, as we shall also see, government investment did increase between 1982 and 1985, overall investment, both public and private, as a share of national income, was not increasing. Indeed our argument has been that PASOK's failure to bring the net PSBR under control was as much due to consumption expenditure as well as the failure to introduce a fairer taxation system, as it was to any great investment drive. It is important to note that this is not a traditional 'crowding out' argument (private investment being crowded out by government expenditure and associated higher interest rates) but that in aggregate for the Greek economy as a whole national expenditure was skewed towards consumption rather than investment.

The question arises that if public deficits were still too high, even under the operation of stabilisation through development, as we have argued, then too high relative to what? It is not clear that economic theory provides any unambiguous conclusions on the appropriate level of public deficits. Setting aside Ricardian equivalence theorems, most of the literature addresses the issue of movements in public deficits over the business cycle and the requirement that the ratio of debt to GDP should be kept constant at mid-cycle (see Spaventa, 1988, p2; Odling-Smee and Riley, 1985). What we can say is that in Greece the level of public deficits in this period was high relative both to historical levels and to other OECD countries.

As Spaventa argues, a long period of high public deficits suggests that fiscal surpluses should be achieved sometime in the future (ie the government has an intertemporal budget constraint). PASOK's inability to increase the tax base, reduce tax evasion and thereby increase tax revenues suggests that there may be social and political limits in Greece to providing future fiscal surpluses. This has clear implications for the government's ability to continue servicing the debt. At some point public confidence may be undermined (leading, for example, to increased risk premia exacerbating the problem), thereby bringing into question the sustainability of the stabilisation through development strategy. Such considerations probably played some part in the adoption of the more austere stabilisation approach in 1985.

We would argue that the ambiguity over PASOK's policy in this area reflects an underestimation of the degree to which a high net PSBR, especially one which is a result of consumption expenditure, constitutes a problem.[26] As with PASOK's incomes policy, we can say that PASOK's ambiguity with respect to public deficits enabled it to be more loose in the demand-side aspects of its policy than was consistent with its own strategy of gradual adjustment. Furthermore this limited the amount of financial resources available for PASOK's supply-side policies. In this respect, it is interesting to speculate that there may be a level of public deficits which severely handicaps the ability to progress with a strategy of supply-side structural reforms, although this will be taken up more fully in the next chapter.

Table 5.8
Contributions to Growth in GDP 1982-85)

	1982	1983	1984	1985	average yearly contribution 1982-85
Primary Sector	0.3	-1.3	0.9	0.2	0.025
Secondary Sector	-0.7	0.1	0.4	1.1	0.225
Tertiary Sector	1.0	1.6	1.6	1.9	1.525
GDP	0.6	0.4	2.9	3.2	1.775

We further argue that the lack of a demand-side adjustment can be seen if we look in more detail at the macroeconomic picture of the Greek economy in 1985. Firstly, on growth, while there was an improvement, a central question is whether this reflects a new permanent phenomenon - Arsenis (1987, p223) has claimed that on same policy, the growth for 1986 would also have been 3% - and whether there is an underlying improvement in the structural contributions to this growth. Table 5.8 does not provide any clear picture on this.

While the table shows the increasing importance of the service sector to growth (reflected in the rising total size of the service sector in GDP), this would not necessarily contradict the major goals of PASOK since the supply-side policies to restructure and modernise Greek industry could only be expected to bring results over the medium to long term. Although Arsenis claims that manufacturing output did recover in 1984 and 1985, the Bank of Greece reports that this was not enough to cover the fall in the previous three years (Bank of Greece, 1987, p59). And as the OECD concludes over 1984 and 1985:

> In contrast to what could have been expected in a cyclical upturn and to developments in the rest of the OECD area, the recovery in manufacturing was very modest. As a result, its level in 1985 was still about 4% below its 1980 peak, whereas in almost all OECD countries output exceeded its previous peak. Moreover the recovery was narrow and fragile as many important sectors recorded substantial declines in output reflecting structural weaknesses... (OECD, 1986, p10).

We should add that the first PASOK five-year plan (1983-87) was based on an average annual growth forecast of 3% to 4% for the period as a whole. And as Table 5.6 also shows there was no improvement in unemployment, even given the dampening effect of public sector employment policy.

Arsenis (1987, p224) claims that inflation was reduced from 25% in 1981 to 16-17% in July 1985 (Arsenis estimated a 12% inflation figure for 1986 on same policy). However the gap between Greek inflation and that of the EEC was not narrowing in this period, but on the contrary remaining fairly constant. And secondly from Table 5.6, we can see that if we take as our measure of inflation the GDP deflator, as opposed to the consumer price index, then there is little improvement over this period.[27] True, as we have said, Arsenis and Katseli only argued for a gradual reduction in order not to destabilise expectations. But one could argue that since monetary and fiscal targets were not being met, this would have an adverse effect on expectations.

On the balance of payments, a 'successful gradual adjustment' argument would be that there was long-term improvement and that the 1985 figures were adversely affected by the speculative activities already mentioned. However, the figures on Table 5.6 do not show any such improvement, especially if we examine the current account deficit as a percentage of GDP.[28] The picture from Table 5.9 is mixed. It shows that PASOK's economic policy in these years was unfortunately affected by the decline of invisible earnings. There is also some improvement in the amount of imports covered by exports in 1983 and 1984, but not in 1985. 1985 of course was affected by speculative activities and the effects of the October stabilisation measures, something accepted by the Bank of Greece (1986, p21). However the Bank of Greece also estimates that import penetration increased in both 1984 and 1985 (Bank of Greece, 1986, p18) and we have seen that manufacturing production results were modest for both years. Whatever the long-term prospects for the supply-side restructuring of industry, the incomes policy we have described was bound to affect the competitiveness of industry in the short run.

Table 5.9
Balance of Payments and Trade Figures

	1980	1981	1982	1983	1984	1985
Trade Balance						
Xs($m)	4094	4771	4141	4105	4394	4293
Ms($m)	10903	11468	10068	9491	9745	10561
Xs % GDP	10.16	12.81	10.76	11.82	13.41	13.10
Ms % GDP	27.06	30.79	26.17	27.32	29.74	32.23
GDP ($m)	40291	37246	38469	34741	32769	32760
% Ms covered by Xs	37.55	41.60	41.13	43.25	45.09	40.91
Trade Deficit as % GDP	16.90	17.98	15.41	15.50	16.33	19.13
Net Invisible Receipts ($m)	4593	4276	4042	3510	3221	2992
Invisibles % GDP	12.3	11.5	10.5	10.1	9.8	9.1
Current Deficit as % GDP	5.5	6.5	4.9	5.4	6.5	10.0
Oil Balance ($m)	n.a.	-2900	-2130	-1923	-2187	-2354

Figures for imports (Ms) and exports (Xs) are on a settlement basis.

Source: based on own calculations from OECD (1985-6, Table 5) and OECD (1987, Table 7)

The loss of competitiveness can be see from Table 5.10 where the indicator for competitiveness is the real exchange rate of the drachma (relative unit labour costs at common currency - note a fall in the indicator entails an increase in competitiveness). As we have seen the incomes policy was only really tightened up in 1983, with the ATA payment losses being made up in 1984. Spraos (1986) has estimated that the gains in competitiveness of the January 1983 devaluation were thus dissipated within eighteen months.

Table 5.10
Real Exchange Rate of Drachma
(First Quarter 1983 = 100)

	All Countries	change from previous period (%)	EEC countries	change from previous period (%)
1978	84.9	0.9	84.0	0.5
1979	87.8	3.4	85.3	1.5
1980	87.8	0.0	84.4	-1.0
1981	94.1	7.2	93.5	10.8
1982	109.2	16.0	109.2	16.8
1983	104.2	-4.6	104.9	-4.0
1984	111.1	6.6	113.5	8.2
1985	108.9	-2.0	111.7	-1.6

Source: Bank of Greece

The loss of competitiveness is also reflected in the decline of private investments in manufacturing (Table 5.11).

Of course private investment in this period was bound to be affected by the election of Greece's first socialist government. Apart from the general climate of uncertainty, the private sector would point to the effect of price controls (especially between 1981 and 1983) on profits, the rise in wages already discussed and new labour legislation (which PASOK introduced to amend explicitly the previous draconian and anti-labour Law 330).[29] On the other hand, as we have already noted, the decline of private investment predates the election of PASOK in 1981.

Table 5.11
Private Investment in Manufacturing
(constant 1970 prices)

	1981	1982	1983	1984	1985
Billion drs	13.2	12.6	11.8	11.8	10.6
Yearly increase %	-7.0	-4.5	-5.6	-0.1	-10.2
% GDP	3.2	3.0	2.8	2.7	2.4

Source: Ministry of the National Economy; National Accounts

Thus overall, and even taking into account speculation in 1985, the balance of payments deficit continued to be high thereby increasing Greece's foreign debt (Table 5.12).

Table 5.12
Greek Foreign Debt ($m)

	1981	1982	1983	1984	1985
Debt	7,876	9,499	10,562	12,286	15,220
% GDP	21	25	30	37	47

Source: Ministry of National Economy, Quarterly Information Bulletins

The position of the Bank of Greece, as well as some PASOK officials, was that this increasing debt considerably reduced the creditworthiness of the Greek economy in the international financial markets, entailing the danger that Greece would face higher interest rates in the future.[30] PASOK was to use the level of the debt, and the balance of payments deficit, as the major contributing factors for the need for the October 1985 stabilisation programme. Although Arsenis claims that the debt could have been financed and there was no problem of acquiring new loans, a lot clearly depends on how the loans had been and were going to be spent, as Arsenis himself accepts (Arsenis, 1987, p226-9). As Angelopoulos points out this is particularly important given that an increasing amount of loans are needed merely to pay back old debt thus making it even more important that the rest is used to pay for investment rather than consumption (Angelopoulos, 1986, p66).

There is no doubt that the PASOK government did increase public investment in this period. Spraos (1986) estimates from the National Accounts that public non-housing investment rose from an average of 4.6% of GDP in 1979 to 5.4% in 1982-84, while it was 6% in 1984 and higher still in 1985. But as Spraos concludes:

> While, therefore, the rise in public sector investment was in keeping with the ideology professed by the Government, there was nothing unique about the level, particularly in a situation in which some of the investment high ground was being vacated by the private sector.

Given this it is not surprising that the structure of GNP does not show any improvement in this period (Table 5.13).

Table 5.13
Composition of GNP at Market Prices (% shares)

	1970	1980	1985
GNP	100.0	100.0	100.0
Private Consumption	67.1	62.8	65.7
Government Consumption	12.4	16.4	19.9
Fixed Investment	24.0	21.5	18.8

Source: OECD, 1987

The weaknesses of the Greek economy revealed by these figures was increasingly accepted by PASOK spokesmen after the 1985 election. Papandreou (1986 p8) stressed the size of the balance of payments deficit in 1985, and V Papandreou (1986a) in a parliamentary speech claimed that this necessitated an EEC loan. Simitis (1986, p18) claimed that the economic developments that led to the October 1985 stabilisation measures revealed the extent to which imports increased even with limited increases in national income, reflecting the weakness of national production and its poor competitiveness, and the extent to which the Greek balance of payments still relied on invisibles. Papantoniou (1987b) claimed that the high level of public deficits was, and had been, a major obstacle to the resources available for development. Thus the macroeconomic record up to 1985 was presented as the rationale for the introduction of the stabilisation programme. We have seen in this chapter that the macroeconomic record was indeed a poor one, and our argument has been that PASOK did not in actual fact implement the degree of control implied by its strategy of 'gradual adjustment'. We thus need to conclude by examining this phenomenon in a little more detail, in particular to see if we can reach conclusions which are of more general relevance.

5.5 Conclusions

An overall assessment will have to await a careful examination of some of PASOK's structural policies, aimed at setting the foundations of the supply-side response. However, some preliminary and tentative conclusions can be drawn here.

Firstly, we have seen that PASOK economists argued that the

policy of 'stabilisation through development' implied a complex and coordinated package of supply-side and demand management measures. Our analysis here would suggest that at least on the demand management side, the degree of coordination and control entailed by the idea of gradual adjustment was lacking. Indeed we have noted that there was some ambiguity over whether PASOK's strategy entailed reducing the net PSBR as such or merely restructuring public expenditure towards investment. We would argue that this ambiguity reflects an underestimation of the effect of public deficits on general economic stability, and the degree to which they impose limits on the resources available for the various supply-side policies to be discussed in subsequent chapters.

We have noted, in particular, the failure to meet monetary and fiscal targets in either 1984 or 1985. It may be thought to be no coincidence that 1983, where targets were more or less reached, was the only non-election year (in 1984 there were elections to the European Parliament which were turned into a referendum on the government and in 1985 there were national elections). The existence of an electoral business cycle is hardly evidence of a consistent application of gradual adjustment. Thus in 1985 not only did public deficits shoot up but public sector employment, including the DEKO and banks, increased by 6.5%. The parallels with a similar phenomenon in 1981, under New Democracy, are uncomfortably close (see OECD, 1982 for overshooting of monetary and fiscal targets in 1981). Nor can it be said that yearly macroeconomic policy was closely related to the five-year plan to enhance coordination of macroeconomic and supply-side policies, as was envisaged by Korliras and Katseli. Not only was the plan based on higher growth rates than actually realised but its relevance can be gauged from the fact that the 1983-87 plan was discussed in Parliament only in 1984.

Incomes policy was also not integrated consistently into the overall strategy. Indeed the increases in wages, especially those of 1982, were defended by PASOK spokesmen, both before and after October 1985, on the grounds of social justice. While such a consideration is clearly important for a socialist government, we have suggested that the short-term consequences of wage rises, especially for industry, were probably underestimated.[31]

It is worth mentioning that PASOK did not develop a kind of incomes policy discussed in chapter 3.6 to 3.7. There was hardly any attempt to reach a 'social contract' with unions and the Government relied on its power to enforce its will through the arbitration process and in 1983 even went as far as to make its incomes policy legally binding. While part of the problem is no doubt the weakness and limited autonomy of Greek trade unions, PASOK did not seriously attempt to change this 'given' fact. For all PASOK's repeated declarations of the need for consensus and dialogue as prerequisites for the success of its strategy, this never extended to discussion or negotiation over its macroeconomic strategy.

The reasons behind the lack of coordinated and consistent application of policy are obviously numerous and complex. So far we have assumed a united party attempting to impose a relatively coherent package of measures on a certain economy. In the 'real' world a party is seldom so united on one policy that it is determined to implement it come what may. Thus the macroeconomic record of these years bears the imprint of various political and economic pressures. What perhaps distinguishes PASOK more is the poverty of its internal institutions for developing policy which harmonises the various social interests and political opinions that it represented - suffice it to say here that the first party Congress was held only in 1984. Lacking the mechanisms and institutions to formulate a 'global' strategy, that is one which attempts to harmonise various interests into a strategy which most can feel some commitment to, various social, economic and political interests and pressures represented by PASOK had to find alternative means of expression.

We have already noted the difficulties entailed in distinguishing policy actually carried out from the policy which was pronounced. Thus after its first election victory in 1981 PASOK, besides implementing its economic strategy, had various other objectives. One was clearly building up its social base, creating the conditions for its continued hold on government. The inability to meet certain economic targets can be seen in this light. The use of public sector employment as social policy, the inability to tackle radically the problems of tax evasion and expanding the tax base, and the various credits and implicit subsidies to farmers and other social groups all contributed to the failure of controlling public deficits, especially in a context of attempting to increase public investments. But these also all reflect, in one way or another, an attempt to build up PASOK's social base. Increasing taxes or failing to employ people in the public sector (especially in Greece where such a practice has such a long tradition) threatens to alienate important interest groups.

For PASOK this was particularly serious given that it appealed to a wide variety of social groups (the 'alliance of non-privileged Greeks' - chapter 4) but more importantly, since nearly all parties seek to build social coalitions, it had not attempted to harmonise the conflicting interests of such groups. This is true both in the sense of inter-group differences (farmers and urban industrial workers, say) and of conflicts between short-term and long-term interests.

For a new party in power does not act upon a tabula rasa. Lacking a clear global strategy to mediate interests, to set priorities and provide ministries with operational plans, pre-existing and powerful pressures begin to come into play. This process is not only reflected in the failure to deal with public deficits but also in the 'style' of the PASOK government. Dretakis' tax on property was watered down after 'pressure' was put on Papandreou concerning the unpopularity of the scheme. If the leading economic officials were committed to 'stabilisation through development', then other sections of PASOK could work to build up the social base by employing more people in the public sector. This lack of coordination in economic policy was severely criticised by Arsenis himself, after October 1985. He also notes the difficulty he had of generating serious and transparent discussion of policy within PASOK (Arsenis, 1987; Koulouglou, 1980). The lack of an incomes policy, in the sense of chapter 3, is also important in this respect. For it allowed a more redistributive incomes policy than might otherwise have been the case and gave PASOK a freer hand, more autonomy, to adjust incomes according to specific political considerations of the moment.

By 1985, while the extent of the economic 'crisis' can be debated, it is clear that macroeconomic imbalances remained very severe. In chapter 3.5 we argued that the coherence of any alternative economic strategy depended, in part, on the extent to which macroeconomic control was exercised in order to provide a favourable framework for supply-side policies and to facilitate the government's ability to make critical decisions on economic priorities. The failure of 'gradual adjustment' suggests that PASOK's policy implied a degree of coordination and systematic application that was simply lacking. This further implies that the supply-side response, to which we now turn, would have to be considerable if the coherence of PASOK's alternative economic strategy as a whole was to be maintained.

Notes

1. The OECD estimates that the service sector accounted for about three-quarters of the rise in GDP after 1975 compared with a little over half before that (OECD, 1982, p8). On the fall of non-housing fixed investment as a percentage of GDP see OECD, 1982, p12.

2. Giannitsis gives a detailed analysis of the worsening structure of Greek industry in 1970s, and the return to traditional activities (Giannitsis, 1986; OECD, 1982, p10). The OECD also estimates that the annual rate of growth of manufacturing output of 4% during 1975-81 was less than half the earlier period. These issues are discussed at greater length in chapter 7.1.

3. A falling trend in all three categories after 1981 was to have serious implications for PASOK's macroeconomic policy (see Table 5.9).

4. Expansionary policies, unaccompanied by structural interventions, led to an accumulation of inflationary pressures:

    something which explains to a great degree why the shock to the Greek economy was more strong after the second oil crisis of 1979. But even after 1979 there was no consistent implementation of a policy of stabilisation and adjustment of the economy to the new conditions. In 1981, especially, an excessively expansionary fiscal policy was carried out, which manifested itself in an increase in the borrowing requirement of the public sector, as a percentage of GDP, from 8.4% in 1980 to 14.8% in 1981. (Bank of Greece, 1986, p21).

5. The existence of an expansionary monetary policy since 1975 has been questioned by some observers, notable Vergopoulos. In his comparative study of Spain, Greece and Portugal, Vergopoulos (1986) argues that in Greece ever since 1974 successive governments have operated a contractionary monetary policy, and that in the Greek context inflation is associated not with an expansionary

policy but with a contractionary one. Unfortunately Vergopoulos uses M2 as his measurement for the money supply and M1 as a percentage of GDP as a proxy for the degree of liquidity. Had he used the broader M3, which includes bank account holdings, his conclusions on the existence of a contractionary policy could not have been sustained. For M3/GDP is a far more satisfactory proxy for the state of liquidity in the economy (for figures for the growth of M3, see Bank of Greece, 1986, p80-85).

6. The Greek system of industrial relations is characterised both by strong state intervention and by relatively weak trade unions, especially in the private sector. Unsuccessful collective agreements lead to a system of arbitration in which the state has the power to enforce its guidelines. Arbitration is undertaken by administrative arbitration tribunals, chaired by a judge and: 'Since the judge usually (in practice, without exception) follows the government guidelines, the government - through its representative - can control the outcome by supporting either the workers' or the employers' position' (Sambethai, 1986, which includes an excellent analysis of the context of wage regulation in Greece). PASOK's guidelines in 1982 were legally imposed on permanent civil servants and through collective agreements/arbitration tribunal decisions for other workers.

7. Sambethai has estimated that this scale entailed a 35% rise for low (minimum) wages, a 10% rise for average wages, and no rise for higher wages (Sambethai, 1986, p9).

8. There is some controversy on exactly how this incomes policy was imposed in practice. Some private companies seem to have worked out increases based not on actual incomes but on minimum wages, thus mitigating the redistributive effect (see Kapsis, 1983). On the other hand there was wage drift in the private sector and some public employees (especially bank employees after a six week strike) received considerably more than the 'corrective amount'.

9. As we shall see later this deterioration is not expressed in the balance of payments improvement for 1982, which was mainly due to a reduction of petroleum stocks.

10. See also OECD (1983) pp44-6.

11. Korliras also pointed out that the speed of approaching the desired growth path will also depend on such factors as world economic developments, and that the 'medium-term stabilisation and recovery program ought to correspond to a rolling five-year plan' (Korliras, 1986, p38).

12. It was pointed out that the existence in Greece of a large number of small-medium firms in the manufacturing and handicraft industries was accompanied with a reliance on the intensity of labour in production rather than the efficient use of capital, new investments or organisational restructuring. Thus even if in the long run the solution lay in more investment and structural changes, in the short run wage increases could have a powerful effect on competitiveness (Commercial Bank of Greece, 1983, p14-15).

13. There is frequently a confusion of how such measures help to improve the balance of payments. The best case scenario is that reducing costs of production leading to increased investments and improved competitiveness promoting exports - although in this case it is not entirely clear how realistic this could be, given that producers knew that costs would increase again in the near future. Often, therefore, the dominant mechanism is that real wage reductions by reducing demand also reduce imports.

14. The first tranche was adjusted from 35,000drs to 50,000drs, the second from 55,000drs to 75,000drs and the third from 80,000drs to 100,000drs.

15. See for instance the report of the PASOK working team on the proposed 1984 budget, claiming that the strategy was both to reduce the net PSBR and to ensure an increase in the share of investment in total public expenditure (PASOK Working Team, 1984). However in an interview

in 1984, Arsenis seemed to suggest that decreasing public deficits as a share of GDP was not as important as restructuring public expenditure towards investment (Arsenis, 1984). In another interview in December 1984, Arsenis made it clear that the 1985 Budget was also to be characterised by a restrictionary outlook (quoted in Giaxnis, 1985). However in a personal interview (in March 1988), Katseli expressed the view that reducing the public deficit, as such, had never been the policy of PASOK. Rather the policy was to restructure public expenditure towards investment.

16. Spraos (1986) estimates that the combined social expenditures of the government and public insurance funds rose to 18.9% of GDP in 1982, from only 15.5% in 1981. He quotes the study by Paulopoulos estimating the corresponding figures for 1983 and 1984 as 19.5% and 19.7%.

17. The OECD has estimated that a reduction in tax evasion could bring in as much as 4% of GDP, while income tax on relatively more affluent farmers could bring in a similar amount.

18. After the October 1985 stabilisation programme both came to oppose what they saw as the shift in policy which now stressed an orthodox stabilisation approach, rather than the previous stabilisation through development.

19. Arsenis estimates that inflation by July 1985 was 16% - the figure for 1985 of 23% of course takes in the effect of the stabilisation measures of October and thus can be misleading.

20. Monetary policy in the 1970s and 1980s tended to be accommodating in Greece, basically because of the need to finance public deficits. Before July 1985 the government was reluctant, or unable, to sell Treasury Bills to the general public and to pay market rates (there is in any case a limited non-bank financial sector in Greece). Thus the inability to sell bonds to the public creates severe difficulties in controlling M3. On the overshooting of monetary targets mainly to finance the

PSBR, see Bank of Greece (1986, p9). However, we should note that whilst monetary targets were not being met in this period, since 1982 important reforms have been introduced into the operation of the monetary system. Interest rates have been increased, limiting the previous implicit subsidisation, interest rates have been simplified and the number reduced (in an attempt to limit round-tripping) etc (see Bank of Greece reports and Halikias, 1987).

21. Indeed Arsenis implicitly accepted that targets were not being met. He recounts that in the Autumn of 1984 he had argued for early elections: 'I personally believed that this [early elections] was the only way we could retighten our economic policy, which had become more loose than it should have, and to change the political balance of power to progress in the field of the problematic industries and the modernisation of the banking system.' (Arsenis, 1987, p202).

22. It should be noted that employment in the public sector has long been in Greece, mediated by a form of clientelistic politics, an important aspect of the social policy of all Greek governments (see Tsoukalas, 1986; Mouzelis, 1978). Thus the OECD estimates that employment in government increased at an annual rate of 5% during the 1970s, while employment in the seven largest public corporations increased by almost one-half in the eight years up to 1978 (OECD, 1983, p24).

23. The most significant contribution to DEKO deficits was made by the social security fund (SSF) and the farmers insurance organisation (FIO). In 1982 the deficits of the SSF and FIO rose to 27.2 and 16.3 billion drachmas respectively representing 66.5% of the total deficit of the 20 major DEKO (Commercial Bank of Greece, 1985, p4).

24. Dretakis' attempt to bring in a wealth/property tax in 1982 to combat the black economy and tax evasion (especially with respect to house building, construction and self-employment) was significantly watered down after Papandreou, responding to pressure (from his own supporters as well as opponents) intervened (see Dretakis,

1987; Arsenis, 1987). Giaxnis notes that whereas Arsenis was supposed to introduce a package of measures to improve and modernise the taxation system, he limited himself to announcing tax reductions and general statements on the need to combat tax evasion (Giaxnis, 1985; see also Agapitos, 1986).

25. It is well known that it is difficult to deduce causation from such national income identities. However, we believe that this analysis is useful for here and elsewhere in this chapter we have argued that part of PASOK's strategy entailed using fiscal policy as a policy lever. Since PASOK failed either to reduce the net PSBR or to significantly restructure government expenditure towards investment, for reasons discussed in this chapter, the analysis here seems appropriate.

26. Odling-Smee and Riley (1985) provide a useful discussion of the need for government borrowing to finance investment rather than consumption expenditure.

27. There are some good reasons why this in fact is a better measurement of inflation, mainly because it includes the prices of all goods that make up GNP, not just consumer goods.

28. There are good reasons why this is a better indicator. For instance between 1980 and 1984, the US dollar was appreciating so that a given deficit in nominal terms was increasing in real terms. Furthermore the improvements of 1982 for the current account (-1.9 billion, Table 5.6) is not a true reflection, given that stocks of petroleum were run down to make up for the accumulation of stocks in 1981 (see Bank of Greece, 1983, p14).

29. For a strong critique, from the right, of the effects of PASOK's policy on the private sector, see Drakos (1986) and Kolmer (1986).

30. This opinion was expressed in interviews (in March 1988) with PASOK officials and the Governor of the Bank of Greece.

31. The PASOK argument becomes even more curious after October 1985, since as Table 5.4 shows, the effect of the 1985 stabilisation measures was to reduce average earnings below their 1981 level (although some greater equality within the wage earnings class had been achieved as the result of differentials being squeezed).

6 Democratic planning, the public sector and socialisation

In chapter five we looked at PASOK's strategy of 'stabilisation through development' which rested on an attempt to respond to Greece's macroeconomic imbalances and longer-run structural problems by a coordinated mix of demand and supply-side measures. In this, and the following, chapter we examine PASOK's supply-side strategy. In chapter four we saw that PASOK argued that, given the large number of structural problems in the Greek economy, the role of the public sector was of paramount importance if Greece was to be set on a path of self-sustaining, though not autarkic, development. This entailed a system of democratic planning to set up basic social and economic priorities, 'active' planning to promote the dynamism of the public sector and harness the private sector to national development priorities, and an increase in public investments to begin to reverse the stagnation in investment of the 1970s. This package of measures constitutes the alternative economic strategy of PASOK and can usefully be compared to our planning schema in chapter 3.8.

Of course the match between PASOK's conception and that of chapter three is not a perfect one. This reflects the fact that planning mechanisms and institutions will have to respond, in any concrete case, to the political economy of the country concerned. Thus our discussion here will not cover all the elements of chapter three. For instance, the need for a National Investment Bank reflects British debates and the lack of such an institution, whereas Greece not only has two such state development banks but the state has significant control over financial institutions. Nevertheless our discussion in chapters six and seven constitutes an essential test case for the institutions and mechanisms of interventionist planning. Our goal is to shed light on both the relevance of the left social democratic model for a country like Greece and on what can be learnt from the Greek experiment on the coherence of the model itself.

In section 6.1, we look at the preparation, coherence and implementation of PASOK's first five-year plan (1983-87). In section 6.2, we examine the role played by public sector investments in the period of the plan and look at some of the instruments employed by PASOK in order to promote 'active' planning. We concentrate on planning agreements, public procurement policy and incentives to industry.

In section 6.3, we take as our case study the policy of 'socialisation' for the public sector nationalised industries. This strategy provides a crucial test case for both PASOK's institutional interventions and its ability to integrate public sector firms into the proposed planning mechanisms.

6.1 Democratic planning

Greece has had a number of five-year plans since the 1950s but none have played any decisive role in determining the economic policy to be implemented (Perrakis, 1985; Katseli, 1986). For PASOK, however, the preparation and implementation of a five-year plan was an integral part of its alternative economic strategy (chapter 4.2.2). The provisional version of the 1983-87 plan stated that it was '... an instrument aiming at the co-ordination of economic and social policy, at the mobilisation of the production possibilities of the country, and the promotion of self-sustained economic development'.

The correspondence of PASOK's conception with that discussed in chapter 3.1 is a close one. Firstly, the plan attempted to link economic and social goals, aiming at: 'a total and simultaneous development of the economy, society, technology, environment, education and culture..... An essential feature of the plan is that it combines together economic, social and cultural targets and measures and places equal emphasis on both quantitative and qualitative development indicators...' (quoted in Perrakis, 1985, p12).

Secondly, PASOK argued that it intended to promote a number of novel features into the planning exercise which would facilitate transforming these general social goals into reality. Three elements, all discussed in chapter 4.2.2, were the radical expansion of participatory procedures for both the preparation and the supervision of the plan; the emphasis on setting up a planning process and introducing various institutional measures to facilitate planning; and the combination of indicative aspects of planning with more 'active' planning measures. Thus PASOK's approach can be seen as a test case of what is involved in any left social democratic attempt to go beyond indicative planning.

6.1.1 The 1983-87 plan process

Overall responsibility for the preparation and implementation of the plan lay with the Ministry of National Economy (MNE) while the technical support aspects of the plan were undertaken by the Centre for Planning and Economic Research (KEPE). Both had to co-operate with the public sector (central administration, economic ministries and nationalised industries). The decentralised, or regional, input was mainly from the Nomos councils (Greece is administratively divided into 52 Nomoi or prefectures). These councils were chaired by the (appointed) Governor of the Nomos and included elected representatives from the local authorities, employer and professional associations, trade unions, etc. Broader popular participation was introduced with Law 1270/81 which instituted citizens meetings at the local level which could suggest proposals and priorities for the preparation of the plan. It was envisaged that these citizens meetings would play some role of social control during the implementation stage.

The preparation of the plan can be divided into three phases: project initiation, evaluation and selection. By 1982 the Nomoi councils were instructed to specify their major development objectives and to provide a ranking of basic development projects. In the project evaluation phase there was an iterative process (essentially done twice) to evaluate the various projects coming from the Nomoi Councils, Ministries and nationalised industries (Katseli, 1986, pp69-70). The restraints of macroeconomic policy and available resources were applied and projects were evaluated in terms of certain economic criteria (see below) and their mutual consistency. This led to the project selection phase where the final responsibility lay with the MNE, since the Nomoi Councils were primarily advisory and supervisory organs. However the basic criteria were: priority ranking by the Nomoi Councils, the effective completion of on-going projects, the satisfaction of basic social needs, the expected net economic benefits, the creation of new employment opportunities, the generation of foreign exchange earnings and the contribution to balanced regional growth.

The 1983-87 provisional plan was approved in November 1983, while the final version was passed only in December 1984, entailing a delay which is a recurring characteristic of various plans in Greece.

6.1.2 The main goals of the 1983-87 plan

At the macroeconomic level growth was forecast at an average of 3-3.5% for the period as a whole, based on a relatively smaller annual rate up to 1985 and faster than the average after that. Unemployment was planned to fall to 4-4.5% by the end of the decade, while inflation was planned to reach OECD levels gradually. Resources were expected to come from three distinct sources: private or net government saving and foreign capital inflows (Katseli, 1986, p71). The plan aimed at a reduction of overall public deficits, restructuring public expenditure towards investment and reducing the balance of payments current account deficit to 3-3.5% of GDP by 1987. Net national savings of both private and public sectors were planned to increase by about 4% as a share of GDP. The above package is of course the macroeconomic side of the 'stabilisation through development' strategy analysed in the previous chapter. Table 6.1 summarises the plan's conception of the sources of finance.

The planned increase in investment is shown in Table 6.2. This increase in investment was the key mechanism, through the annual budgets and public investment programme, ensuring that the plan went beyond a statement of general socio-economic objectives (Katseli, 1984). This initial push from public sector investments, was intended eventually to lead to an increase in private sector activity as a result of improved economic conditions. The sectoral aspects of the plan and industrial policy are discussed in the following chapter. However it should be pointed out that many of the plan's goals were very general indeed. The goals of improving competitiveness, diversifying production and increasing efficiency are ones found in most plans and are of relevance to any time horizon.

Table 6.1
Sources of Finance (as % of GDP at market prices)

Sources of finance	1982	1987 Alternative scenarios	
		3% growth	3.5% growth
Private savings	15.5	16.5	17.5
Depreciation	8.3	8.3	8.3
Current Saving of Government [1]	-4.7	-3.0	-2.0
Net Borrowing and Transfer of capital from abroad [2]	3.8	1.5	2.0
TOTAL	22.9	23.8	25.8

[1] ie General government deficit according to National Accounts
[2] includes borrowing from abroad and transfers mostly from the EEC
Source: KEPE, 1987

The 1983-87 plan included a number of further goals to establish planning mechanisms, such as planning agreements, to promote the decentralisation of power, to encourage a new class of investors, and to build up a planning process and a more favourable framework for development. The plan aimed at increasing the percentage of the public investment programme regionally administered, setting up the new <u>Nomoi</u> Councils and their role in planning, and introducing 'planning contracts' between the public sector and local authorities (MNE, 1986, p2). Laws 1270/82 and 1235/82 sought to increase the responsibility and independence of local authorities.

Furthermore the plan envisaged support for investments from new social organisations such as co-operatives and local authority firms (see Law 1416/84 and 1541/85).[1] The above are characteristic of PASOK's conception of what was entailed in going beyond indicative planning.

Table 6.2
Capital Formation (as % of GDP at market prices)

	1982	1987 Alternative Scenarios	
Capital Formation		3% growth	3.5% growth
Gross Fixed Capital Formation	19.1	20.1	22.5
- Private	(13.2)	(12.4)	(14.0)
- Public	(5.9)	(7.7)	(8.5)
Inventories	3.8	3.2	3.3
TOTAL	22.9	23.3	25.8

Source: KEPE, 1987

6.1.3 Co-ordination and participation in the plan

PASOK argued that the novelty in its approach to planning lay not only in the goals and instruments of the plan but in setting up a planning process. Katseli (1984) mentioned four aspects of this: the organisational framework for planning, an institutional framework for clarifying the role of both private and public sectors, a network for the collection and processing of information and information exchange, and a network for conflict resolution. Indeed in chapter 3.5, we argued that the coherence of the whole alternative planning 'package' rested on a resolution to such problems as conflict resolution, administrative efficiency and information flows. It is to such questions that we now turn.

<u>Plan co-ordination at the level of society</u> In chapter 3.1 we argued that any national plan would need to initiate debate within society on alternative social and economic priorities. Although the unions and various employer and professional organisations had a certain amount of input into the national plan at the local level, through the <u>Nomoi</u> Councils, there was no equivalent at the national level. There was no forum for dialogue and consensus-building during the various phases of

the planning process. This lack was partially rectified in 1985 with the establishment of the National Council for Development (ESAP). As Papandreou put it '...the council constitutes the highest advisory institution for democratic planning in our country and comes to complement, at the national level, the institutional framework for the development and planning of the country' (quoted in Mitrafanis, 1985). ESAP was made up from representatives of the employers and professional associations, the large urban local authorities and the unions.

However, by the time ESAP was established, the political climate for consensus building was not propitious. The implementation of the stabilisation policy in October 1985, and the opposition of the Greek trade union federation (GSEE), meant that there was little room for building consensus on the implementation of the plan. The private sector was equally reticent to indulge in serious negotiation. In any case ESAP was not a body in which votes were taken and PASOK seems to have used it as a 'sounding board' to test public opinion. Controversial aspects of planning, notably planning agreements, were never brought up by PASOK in the ESAP forum. The obstacles to creating institutions for dialogue in the Greek case are discussed later. Here we merely point to two consequences of this phenomenon. Firstly, in this context it was difficult to reach agreement on the central priorities for social and economic development.[2] Secondly, the absence of such a consensus, or dialogue, prevented the plan from becoming the central focus for economic decision-making for either the private or public sectors.

Plan co-ordination at the level of government The implementation of the 1983-87 plan was also hampered by the lack of any government organ responsible for co-ordination and supervision. We have seen in earlier chapters the importance of setting up a strong planning agency, and specifying clear lines of authority between the planning agency and other government departments. Such an agency can also provide a framework for conflict resolution. PASOK had no such new planning agency and there seems to have existed multiple centres of power within the PASOK party and government (Arsenis, 1987, pp75, 146, 184).

PASOK eventually recognised this problem and for the preparation of the 1988-92 plan set up an Executive Council for the Implementation of the Five-Year Plan (ESEP), directly under the MNE. ESEP originally consisted of representatives from four ministries (MNE, National Works and Environment, Interior and Agriculture) and one from KEPE, but was later expanded to a total membership of eight. It was envisaged that ESEP would play a coordinating role while referring any disagreements to the Cabinet for a final decision.

The lack of such a coordinating organ created severe obstacles for the creation of adequate networks for information and information exchange. This was particulary problematic in the relationship between planners and the central institutions. Planners had considerable problems in co-operating with ministries and public corporations and in gathering information and project proposals from them. ESEP later recommended the creation of joint committees of KEPE and these central institutions for future co-operation, a mechanism which would also have enabled central institutions to have an enhanced awareness of the plan and its demand on them.[3]

The relationship of the planners with the regional organs of planning was also subject to problems if slightly less severe. An unpublished memorandum for the first meetings of ESEP noted that there was a '... lack of sufficient direction towards each <u>Nomos</u> on the macroeconomic guidelines and basic directions of the national development policy with which they should mobilise to formulate their proposals and the technical support of the proposals'. To an extent this reflected teething problems of the new regional planning institutions and the necessary learning process involved. ESEP later recommended a number of improvements, including a closer relationship between KEPE and the regional planning organs. However, there is also the question of resources for planning at this level. The <u>Nomoi</u> Councils and citizens meetings clearly needed technical and educational back-up services. Although some progress was made in this area during the 1983-87 plan, there was still a lack of an overall organisational network of support services at all levels.

<u>The effects of the lack of co-ordination</u> A central consequence of this lack of co-ordination in the planning process was the continuation of the customary <u>ad hoc</u> intervention of various ministries. For instance although KEPE is in principle a semi-autonomous institution responsible for the technical aspects of

the plan, it faced continued <u>ad hoc</u> interventions in its operations (see Lambrianidis, 1987, pp31-2).[4] Such a practice prevented KEPE from establishing a 'corporate loyalty' for promoting the planning process. KEPE officials rarely felt that they had the power to argue their case and therefore limited their role to executing their task within narrowly defined limits of political intervention.

This lack of autonomy was also in evidence with respect to the regional/local planning institutions. Local planning was often reduced to an attempt to maximise available financial resources to promote infrastructural projects in particular areas. While the <u>Nomoi</u> Councils did provide some sort of forum for consensus building, it was a rare <u>Nomos</u> Council that was able to integrate social investment priorities into an overall development strategy.

Furthermore, local planning services were overburdened by their daily task of managing resources available to the <u>Nomos</u>, with the result that all too often local plans were formulated merely because there was some political necessity for their existence. What was lacking in local plans was a theory of how to interpret the phenomena under study with the result that plans were often mere forecasts based on the general 'feeling' of the planners or the extension of existing tendencies without being able to delineate new phenomena and trends and thus new areas for investment (Lambrianidis, 1987, p32). Here too, it was difficult for planners to build up a 'corporate loyalty' for planning or an infrastructure for planning on an ongoing basis. This was not helped by the continued practise of appointing planning officials at all levels on party political grounds.

While the plan included certain measures to strengthen local authorities, PASOK did not take the more radical measures it had promised. In particular, it did not introduce the second-tier of local government, or legislate for local authorities to have their own independent financial resources. Both these steps would have had a considerable effect on increasing the autonomy of local government and improved the nature and quality of the local input into planning. In this context it was difficult for the national plan to be integrated into local planning on the lines discussed in chapter 3.2. For the model of local enterprise boards and local planning rests considerably on the autonomy of local planners developing a strategy for their own community. As Spraos (1986) concludes: 'overall, however, decentralisation is proceeding slowly and gets entangled in the traditional failure to distinguish in the political process between the local community and the local party'. This absence of autonomy reflects the lack of a strong civil society within the Greek social formation and the predominance of vertical linkages over horizontal linkages within society as a whole. This is a theme which runs through much of the rest of this book, reflecting the social, or political economy, prerequisites for introducing planning in any given society.

6.1.4 Plan consistency

The lack of co-ordination, and an adequate framework for information flows and conflict resolution was bound to affect the overall consistency of the 1983-87 plan. Although some analysts, involved in the preparation of the plan, (eg Babanasis, 1984 and Katseli, 1986), have claimed that such consistency was achieved by the iteration process previously mentioned, even a cursory examination of the regional aspects of the plan suggests that this met with only partial success. A classic example of this failure is the inclusion of certain agro-industrial projects, for processing olive oil for instance, in many Nomoi instead of targeting this activity for certain Nomoi. Overall the impression of the plan of a long list of desirable social investment projects constitutes one of the main weaknesses of the plan, thereby limiting the resources available for more directly productive investments.

A further problem concerns the scope of the plan and the inclusion of a vast number of goals and instruments. Some of these goals were bound to be achieved merely by the fact that the plan included the intended strategy of the public sector. In our discussion of the nature of plans in chapter 3.5, we saw that it may be inadvisable for a national plan to be too comprehensive, in the sense of covering all regions or all sectors. We noted that establishing a planning process, with adequate resources, and administration was a time-consuming operation which relied on considerable popular support for the exercise. This is backed up by the Greek experience. PASOK's attempt at comprehensiveness obscured the need to delineate priority sectors or economic activities for planning. It drew attention away from the fact that difficult choices had to be made entailing both future benefits but also present costs. The lack of clarity on such issues had important implications for the consistency of the plan, a theme which is developed in the following chapter. This tendency to include every conceivable goal within the plan itself was subsequently criticised by the ESEP which hoped to focus on a narrower range of priorities for the 1988-92 plan.

The issues concerning the consistency of the plan with macroeconomic policy was the subject of chapter five. By 1985 PASOK economists were claiming that the level of public deficits had constituted a severe obstacle to the implementation of the 1983-87 plan (MNE, 1986, pp10-11). It was further argued that these developments had also restricted the resources available for investment (Papantoniou, 1987b). The plan growth targets were particularly problematic. Although the planners worked out a number of alternative growth scenarios, the one finally chosen was on the optimistic side with the overall plan dependent on reaching the growth target. Indeed the growth targets were less a forecast based on the result of the implementation of the plan and more an estimate of the growth necessary to reach the government's target for the reduction of unemployment: it was estimated that for the plan period as a whole an average growth rate of 3% would secure 150,000 to 160,000 new jobs, whereas a growth rate of 4% would raise that number to 195,000. There are in fact strong pressures on planners to choose optimistic growth targets since these have an important effect on expectations. However the problem arises if such targets become unrealistic during the plan period. Once stated it may be difficult to go back on these figures for political reasons with the result that the plan may still be of some use for political purposes but is of little use for operational guidance in economic decision making. Such a process may have affected the course of PASOK's first plan. Overall we can conclude that the macroeconomic policy actually carried out was one of the most serious problems for the consistency of the 1983-87 plan. For the 1988-92 plan, ESEP promoted the idea of suggesting alternative growth scenarios with the plan being operational whatever the growth rate actually achieved.

A final aspect of the consistency of the plan worth considering is the degree to which it rested on popular support and became a central focus for economic decision-making within society. It is perhaps surprising that of all aspects of PASOK's 'alternative economic strategy', the five-year plan created the least debate and controversy - neither the unions not the private sector concentrated their attention on the plan, to criticise or support it, or to debate its limitations and suggest counter proposals.

To conclude, while we have seen that the plan did attempt the establishment of a planning process, this strategy met with certain obstacles. This limited success was due to a number of factors: lack of experience and technical infrastructure, organisational problems and politics. We have also suggested that all three are to a degree socially determined - the lack of debate and consensus-building, reflecting, in part, the lack of horizontal linkages or autonomy within civil society.

6.2 Public investments and 'active' planning

In this section we examine the process by which PASOK intended to ensure the implementation of the 1983-87 plan, and in particular the role of public sector investments to promote the supply-side response of the economy and various planning mechanisms to harness the private sector.

6.2.1 Public investments

The overall budget for investment projects for the 1983-87 plan was forecast to be 1,350 billion drachmas (1982 prices), while, based on macroeconomic forecasts, the plan foresaw financial resources for public investments amounting to around 1,000 billion (MNE, 1986, p22).[5] The public sector's response, and especially the programme of public investments, was quite satisfactory as Table 6.3 indicates (see also Papantoniou, 1987a, p6). These investments refer both to the Public Investment Programme and to the investments of the nationalised industries (usually termed in Greece 'Public Corporations and Entities', hereafter DEKO). The results of the former were particularly impressive, although better in the areas of transport, health and education than in tourism-cultural projects, irrigation and land reclamation (MNE, 1986, pp22-5). We have already noted that these investments, carried out either by the central administration or the nomoi, were too skewed in favour of social investment projects. While no doubt such projects contribute to the efficiency of other investments, there was clearly an imbalance. Indeed this was implicitly recognised and ESEP argued in favour of focusing the attention of future plans on a narrower range of investments.

The DEKO investments were mostly in the areas of energy, telecommunications, transport and irrigation-sewage. A large part of these investments was taken up by the Public Power Corporation (DEH), aiming to reduce Greece's dependence on imported liquid fuels by developing domestic lignite sources. However, as we shall see in section 6.3, the DEKO were never satisfactorily integrated into the 1983-87 plan, resulting in certain problems with respect to the sectoral and regional consistency of the plan - ie questions pertaining to where to invest, what is the effect of investing in area A on the local economy and what on neighbouring area B etc.

Table 6.3
Overall Expenditure of the Public Sector
(in billions of drachmas)

	1983	1984	1985	1983-85
At current prices	242.9	319.4	395.5	957.8
At 1982 constant prices	194.3	214.7	228.2	637.2
% of take-up of yearly credit				
- with respect to plan forecasts (1350)	14.3	15.8	16.8	47.2
- with respect to available resources (1050)	18.5	20.4	21.7	60.7

Source: MNE (1986)

In addition to the above investments there was a set of proposed investment projects in industry and manufacturing which, while included in the plan, were much less integrated into its structure. We saw in chapter 3.2 that the left social democratic model emphasises the importance of state involvement in directly productive activities, by creating NPEs, expanding the activities of nationalised industries or setting up joint ventures with the private sector. The PASOK proposals in fact originated in the early post-junta (1974) years and were supposed to constitute a new wave of investments in advanced sectors to restore the dynamism of the Greek economy. The most notable of these investments included a petrochemical complex, an alumina plant, a stainless steel and ship scrap yard, and a metallurgical industrial complex (see Papagiannakis, 1988[6]). It is beyond our scope here to fully examine the fate of these projects. Some were finally abandoned (ship scrap yard and the metallurgical production complex, METBA) or were reconsidered (the stainless steel project was found to be unrealistic after a reexamination of the original calculations, while the petrochemical plant was still under discussion, well into PASOK's second term of office). Other projects had their activities redirected - thus the AEVAL ammonia production unit shifted activities to introduce natural gas as part of a very large natural gas project which is now under implementation.

The state's ability to get involved in directly productive activities represents a central failure of the 1983-87 plan. The explanation of this phenomenon sheds some light on the prerequisites entailed in the state adopting such a role.

The MNE (1986, p62) report on the progress of the plan notes two main contributing factors for the delays in this field: the lack of acceptable feasibility studies and the inability of ministers and state investment banks to reach final decisions. The former relates to the underdeveloped organisational framework and poverty of the technical back-up services for carrying out such investments in Greece. Feasibility studies were often based on the mere extension of current demand patterns for certain products, with little analysis of the wider economic and sectoral implications of particular projects. And, of course, the more the delay in the final decision for proceeding with a particular investment, the more unsatisfactory the original feasibility study became.

In chapter 3.4, while discussing the role of state investment banks, we pointed to the need for the issues of subsidisation of investment projects to be clearly delineated in any social-cost benefit analysis. This is clearly important if any government is going to be able to rank priority projects. The PASOK administration provided no framework for such an approach. A possible organisational solution is that the government should set the guidelines for the social-cost benefit analysis, allowing the firm to carry out the actual analysis, with perhaps an independent body assessing the viability of the project, after working out the degree of subsidisation deemed necessary.

Under PASOK, these functions were fused together in an intransparent fashion with no clear division of either responsibility or authority. Projects were subject to continual renegotiation and political bargaining in which political, often electoral, economic and social goals were lumped together making a final decision almost impossible (Arsenis, 1987, p99).

Arsenis (1984) refers to this phenomenon when he notes that Greece lacked an appropriate framework for promoting such projects, with dynamism and imagination, in a coordinated manner. In chapter three we noted that models of alternative planning had not given enough serious consideration to the problems of control, both with respect to NPEs and their internal structure, and to how the public administration, or planners, would relate to state firms. It cannot be said that PASOK attempted to develop any strategy to deal with such questions, and despite various announcements on the need to reform the public administration, it never articulated the process by which this was to be achieved.

The failure of the large state interventions in industry also reflect PASOK's inability to articulate an overall model of the type of development it was promoting; the plan by incorporating a vast number of goals failed to clarify any such model or to set out economic priorities, and the corresponding costs of any restructuring. In some cases PASOK progressed with a certain project without a serious analysis that this was indeed a priority area and without working out the costs and benefits.

At times PASOK seemed to be promoting the importance of industrial development based on large capital-intensive projects to introduce a new dynamism. The fact that many of these projects here originated from the post-1974 period raises the question whether PASOK had by 1981 a clear conception on how best to exploit a new readiness to promote state involvement in directly productive investments. These projects were inspired by a wish to set up a set of investments in basic industries, with forward linkages and to act as a complement to the activities of the private sector. The profitability of these projects, even according to some optimistic feasibility studies, suggested negative or marginally positive rates of return, with doubtful consequences for contributing to an improvement in the structure of Greek industry. They seem more relevant for a model of development for the Greece of the 1950s rather than the 1980s.

Perhaps sensing the limitations of such an approach, and lacking an alternative conception of how, and in which areas to promote state involvement in industry, PASOK elsewhere gave support for smaller and less grandiose projects, with the emphasis on supporting the small-medium sized firms. By 1987 matters were complicated further with the resurrection of 5 or

6 large infrastructural projects (Athens metro, Sparta airport, Achelos dam etc). It is difficult to see any pattern from various PASOK pronouncements on development, given that the various approaches could be seen to be contradictory - to what extent, for instance, do the large infrastructural projects promote national industrial/manufacturing projects?[7] This obviously limited PASOK's ability to articulate a coherent and consistent industrial strategy, in which state activities in directly productive investments could have contributed.

To recap, we have here been examining some important prerequisites for an enhanced role for state involvement in industrial/manufacturing activities, which were lacking in the Greek case and which PASOK was unable to overcome. Firstly, we have emphasised the importance of an efficient and flexible public administration, and a coordinated government response. The way in which such an administration would be able to co-operate with state investment banks, to rank economic projects and delineate the issue of subsidisation, is also of paramount importance. Secondly, we have pointed to a pressing need for planning to be based on a clear conception of the model of development being promoted. This, once more, suggests that the plan, rather than seeking to be comprehensive should seek to pinpoint such priority areas, and that planning and state intervention should be geared to meeting the needs of such priorities. In seeking to be comprehensive, PASOK in the end, was reduced to an inability to use state investments in directly productive activities as part of an overall coordinated strategy.

6.2.2 Active planning

Here we examine three aspects of PASOK's 'active' planning: planning agreements, public procurements and financial incentives. In chapter 4.2.2 we noted that PASOK's approach corresponded fairly closely to that discussed in chapter 3.2. Even by 1987 Simitis, Arsenis' successor at the MNE, reaffirmed, in a parliamentary debate on planning agreements, his party's commitment to such an approach. He noted that PASOK's planning did not end with the formulation of a five-year plan but extended to the creation of institutions and procedures which would promote the participation and co-ordination of all sectors (public, private and social) into the plan, thereby strengthening the supply-side response of the Greek economy

(Simitis, 1987).

Simitis also noted, however, that PASOK had not been particularly successful in this in the period of the 1983-87 plan. While the real average rate of increase in public investments for the plan period was 3.5%, the investment activity of the private sector continued to stagnate (see chapter five). The previously mentioned ESEP memorandum also criticised the plan's emphasis on the Public Investment Programme 'without adequately covering the private sector in the direction of its mobilisation', and proposed that the new plan should have a special section on the role of the private sector with proposals for its integration with government and DEKO policy. Here we can examine how PASOK attempted to come to terms with such mechanisms as planning agreements and what it has learnt from that experience.

Planning agreements   Simitis (1987) claimed that the chief cause of the poor private sector response to the plan was that the existing institutional framework and instruments did not allow the co-ordination of the basic sectors of the economy at the mesoeconomic and macroeconomic levels. Thus although planning agreements were introduced during PASOK's first term, they played a strictly limited role in the co-ordination of the national plan. The planning agreements actually carried out concerned almost exclusively the nationalised sector of the economy.[8]

The legislative framework of the first term did not provide the flexibility to expand the planning agreements mechanism and usually necessitated specific and ad hoc further regulations. Similar problems were confronted with the mechanism of planning contracts. These were established between the central and local authorities (Law 1416/84), but were limited to regional issues rather than to sectoral/industrial ones. There is an important point of general relevance here. A government introducing a new planning alternative, with various new planning mechanisms and institutions, is likely to have to undertake a lengthy legislative programme. The plan, and thus the estimated supply-side response, must be based on a realistic appraisal of what can be achieved in the time available.

Similar arguments are also relevant to the establishment of the necessary co-ordination, administration and technical support services for such new planning mechanisms. Here too

PASOK, at least in its first term, failed to establish the necessary institutions. New legislative measures of PASOK on planning agreements and development contracts, which did constitute a distinct improvement on the previous state of affairs only appeared in 1987. These measures too were never implemented for by 1988 all pretensions that PASOK was implementing an alternative economic strategy had been dropped. However, we conclude by briefly describing the 1987 proposals because they do provide some recognition of the problems of control (which we have discussed here in chapter 3.5), and because they perhaps indicated what should have been done earlier.

The new framework allowed for the first time a dialogue with representatives from the productive classes to ensure a degree of consensus in the operation of these mechanisms, although strangely, these were not brought before ESAP, perhaps because the opposition of the private sector minimised the possibility of developing such consensus. The new legislative framework also sought better co-ordination of planning agreements at the government level. The Committee of Planning Agreements and Developing Contracts (EPSAS) was to co-operate with the MNE, and other economic ministries, to reach planning agreements at the sectoral level. This mesoeconomic co-ordination would be complemented by a micro level framework for development contracts which would be the responsibility of the relevant ministries in co-operation with individual productive units. This approach could in principle have operated on the lines discussed in chapter 3.1 with strategic planning at both the sectoral and enterprise level. Packages of support could have been provided for particular sectors or enterprises, willing to redirect their activities on the lines negotiated with EPSAS. This in turn would have depended on PASOK's ability to articulate a sectoral industrial strategy, which is discussed in the next chapter. But as it was, planning agreements played no role in PASOK's first term of office (1981-85).

Public procurements The ability of the state to carry out planning agreements, as we argued in chapter 3.2.3, depends on the public sector planning its future needs and contracts. We further argued that the approach of using public purchases as an instrument of planning involves a considerable task of administration and information gathering across many ministries, regions and nationalised industries.

PASOK's policy in this area was to use public procurements to support Greek industries/manufacturing activities, thereby promoting and diversifying the productive base of the country, and enhancing the technological and qualitative upgrading of Greek production (Papandreou, 1985a). Public procurements were also intended to encourage a new class of investors (joint ventures, co-operatives, local authority enterprises etc). Papandreou further stated that the strategy would promote increased transparency in the procedures of public procurements entailing reform of the public administration so that central state institutions could coordinate and plan their needs.

The first moves for developing a consistent policy for public procurements were undertaken by Vaitsos at the MNE during PASOK's first term. He set up a Secretariat for Public Procurements with the task of creating the necessary organisational infrastructure and technical know-how. However, before this stage had been completed, the responsibility for public procurements was divided among two other ministries, with the Secretariat moving to the Ministry of Industry.[9] The role of the Secretariat was considerably downgraded in this process, and it was only in September 1987, when the Secretariat had moved once more - this time to the Ministry of Trade - that the new draft of legislation was presented to codify some of the goals of the policy (see the report of <u>Eleutherotypia</u>, 2.9.87). Once more, therefore, progress in this area reproduces the picture of a lack of co-ordination and spasmodic interventions.[10]

<u>Financial incentives</u> Here we concentrate on PASOK's new Law 1262/82 on financial incentives in order to discuss some of the problems associated with the use of such incentives as a complement to national planning (see chapter 3.2.3).[11] Papandreou in 1985 expressed the view that the novelty of PASOK's approach to incentives rested on providing incentives to those firms which could orientate themselves to changing the structure of the Greek economy in an anti-dependency and anti-monopoly direction, as well as promoting a whole new class of investors (co-operatives, local authority enterprises etc).

Indeed there are a number of novelties in Law 1262/82, which replaced the previous legislation (Law 1116/81). Firstly, it relied more on direct grants rather than the interest-free loans, tax incentives and depreciation allowances of the previous legislation. In part this reflected the influence of the EC - about half of all Law 1262 incentives were paid from EC funds.

Secondly, the legislation attempted to promote state involvement in directly productive activities by ensuring that after a certain level (initially 400, eventually raised to 800 million drachmas) the subsidy was matched by the acquisition of an equal value of shares in that company. In the event this aspect of the legislation became largely irrelevant because of private sector opposition and the small nature of those investments carried out. However these are not the only reasons and it is worth reflecting a little further on this phenomenon. In chapter 3.2.1 we noted a number of reasons why an interventionist planning administration may seek such an extension of social ownership - to encourage investment in high technology or high-risk activities, to ensure technology diffusion, to acquire information necessary for sectoral planning and soon. We also noted that the government would have to analyse and develop a strategy where the rationale for such an intervention, and the sectors or enterprises in which it is relevant, is clearly specified. This would have the further consequence of clarifying the state's relationship with the private sector, thereby reducing the potential conflict with the private sector. But we have seen that PASOK was unable, within the framework of the national plan to delineate such clear economic priorities. This may largely account for the failure of PASOK to extend social ownership, within the framework of Law 1262/82 - it lacked a clear conception of where such social ownership could aid its overall strategy of promoting firms orientated to changing the structure of the Greek economy.

A third novel feature of Law 1262 is that it initiated procedures for determining the percentage of subsidy in order to promote government policy. A point systems for assessing projects was initiated which favoured decentralisation, employment creation, utilisation of domestic resources, and vertical integration of Greek production and exports. Points could also be achieved if the project was to be carried out by co-operative or other social ownership enterprises. This new framework had two beneficial results. It encouraged investors to prepare the type of technical economic plans, before starting their investment, which had previously been lacking. It also promoted decentralisation in project evaluation by setting up ten regional advisory committees, which were in a much better position to judge the local implications of the proposed project.

This approach of using a points system to promote government objectives met with only partial success. From MNE figures, it is clear that a large share of the subsidies went to traditional activities and sectors, such as tourism or manufacturing activities with little potential for dynamic growth (see also Konstantinidis, To Bima, 12.4.87). This suggests that risks were not taken, the existing structure of the economy was reproduced and the legislation did not promote the type of dynamism in the economy that was intended. Indeed at times the legislation could work perversely. Thus points could be achieved if a project entailed risk-taking or promoted dynamism, but could also be gained if it was to be undertaken by a co-operative or to be set up in a deprived area. The fact that a particular investment in a traditional sector or activity is undertaken by a co-operative in a deprived sector does not necessarily contribute to the dynamism and restructuring of local economy.[12] The dilemma involved here results from the problem of having too many objectives associated with one planning mechanism. In the case of PASOK's Law 1262 the goals of promoting a new class of investors and economic dynamism were often in contradiction, posing the question of whether the use of incentives legislation is the most appropriate mechanism for encouraging a wide range of social ownership forms within the economy.

However one can say that PASOK did improve the institutional framework for incentives for small investment projects, although the figures for the completion of subsidised projects were not impressive. From MNE figures we have estimated that by February 1988 of the 9,598 investment projects accepted under the legislation only 3,540 were implemented or were in the process of being implemented. Of the investment project completed, the investments amounted to 461 billion drachmas (with a level of subsidy at 145.4 billion) with the creation of 15,000 new jobs. But such figures do not reflect only the inadequacy of the incentives law.

One contributing factor was that investors lacked, or were unwilling to risk, their own capital in projects and that many investors started up the processes of applying for a grant on a 'try it and see' basis. Furthermore the private sector argued that whatever the merits of the new legislation, the existence of a number of other 'counter incentives' made most potential investment projects unrealistic.[13]

The most important point to be made here is that any incentives law can only operate within a context of a favourable environment for investment: '...incentives might promote investment and subsequent growth, but in no case will they be the crucial factor in planning and in realising an upward trend in private investments' (Xanthakis, 1985, p18). So while Law 1262 was an important new mechanism, we would argue that the poor response from investors was only very partially due to some of the inadequacies of the law. We return to this question when discussing PASOK's industrial strategy in the following chapter.

6.3 Planning, socialisation and the DEKO

As the discussion in chapter three made clear, the ability of a socialist government to integrate the public sector into its overall economic strategy constitutes a basic prerequisite for the implementation of planning as a whole. Furthermore we noted that the mere fact that certain enterprises were state-owned did not entail that governments have found it particularly easy to successfully integrate them into the planning process. In particular, we pointed to a set of problems of integrating public enterprises into planning under the heading of control (chapter 3.2).

The ability of PASOK to integrate public firms into its planning alternative is the subject of this section. The general principles of the policy of socialisation for the public sector - in particular for the three largest enterprises involved in energy, telecommunications and transport, respectively DEH, OTE and OSE - were laid out in chapter 4.2.1. Socialisation here has a specific meaning referring only to the policy for the existing Public Corporations and Entities (DEKO) and should be distinguished from the wider, and more general, usage of the term.

The socialisation policy had two aims. On the one hand, the policy was part of PASOK's ideological commitment to promote institutional and structural reforms in order to expand participation in economic decision-making and introduce elements of social control within the public sector. On the other hand, PASOK realised that for its new institutional measures to succeed, and thus achieve a greater legitimacy, they must also work in favour of reconstructing and modernising the DEKO. Socialisation was both an end in itself and a means for economic efficiency and reconstruction. Therefore an examination of the course of the socialisation policy is not only an important test case for PASOK's overall approach. It can help to reveal the problems confronted by such a conception when implemented, but also shed light on certain prerequisites necessary for any strategy aimed at the same time of expanding participation and social control and integrating public firms into the national plan and thus promoting their economic reconstruction.

6.3.1 Economic and organisational problems of the DEKO

The poor economic performance of the DEKO since the late 1970s presented a serious obstacle to the implementation of PASOK's strategy, which relied heavily on public sector investments enhancing the supply-side response of the economy. The combined deficits of the DEKO rose from 2% of GNP in 1979 to 5.5% in 1985 of which 2.8% was covered by government transfers (OECD, 1987, p43). These aggregate figures include the increasing contribution to the deficit, and thus to the macroeconomic imbalances, of those DEKO concerned with social security pensions and insurance (see chapter 5). Of the three DEKO with which we are most concerned here, the financial performance is more mixed, although that of the Greek Railway Organisation (OSE) is the most problematic.

Table 6.4 gives an overall picture of the financial and investment performance of the most important DEKO for our purposes. Although investments of the DEKO were to have an increasing share of overall investments in PASOK's first term, the scale of the task confronting PASOK is indicated by the fact that: 'plagued by high operating deficits and cash-flow problems, investment by the DEKO has remained insufficient, thus retarding the modernisation of infrastructure. Indeed, in many areas (eg transport and telecommunications) public services are clearly inadequate for a country at Greece's level of development' (OECD, 1987, p46).

Table 6.4
Selected Data on DEKO

	1980	1982	1985
Total Employment	116,130	133,995	144,001
of which:			
Telecommunications organisations	30,236	31,148	30,571
Public Power Corporation	26,365	26,673	27,633
Railways	12,905	14,014	14,965
Urban Transport	12,907	13,371	14,275
Olympic Airways	7,869	8,735	11,104
Post Office	9,248	9,355	10,405
Water and Sewage Authorities	3,747	4,272	4,816

Fixed Investment (National Accounts Basis):			
% of total fixed investment	16.5	22.5	30.2
% of GNP	3.0	3.1	3.3

	1980	1983	1985
Administrative Basis (as % GNP):			
- Operating deficit [1]	0.7	0.8	1.9
- Capital deficit [2]	2.0	3.0	3.6
- Total deficit	2.7	3.8	5.5
- Official transfer/grants		2.0	2.8
- Total borrowing		1.8	2.7
Cash Basis (as % GNP):			
- DEKO borrowing requirement	1.9	2.8	2.2

[1] Gross revenue minus material costs, wages and salaries, interest etc.
[2] Investment less capital revenues (including amortization).
Source: OECD (1987)

We can delineate three main factors underlying this poor economic performance:
  a) The DEKO lacked a clear state policy for their direction and development. We should add the traditional lack of autonomy of these DEKO and the existence of <u>ad hoc</u> ministerial intervention at all levels of their operation, as well as the use

of the DEKO for social and anti-inflationary goals without covering this with equivalent sums from the state budget (OECD, 1987; CBG, 1985).

b) A lack of a satisfactory system of corporate planning, uncoordinated investments, waste of resources, insufficient study of costs and benefits of particular projects etc.

c) The poor organisational structure of most DEKO, the effect of a long history of clientelistic politics in recruitment and the non-developmental role that the public administration has played in Greece since the nineteenth century. This has led to low productivity, poor utilisation of manpower and low quality of services produced.

6.3.2 The socialisation framework of Law 1365/83

There was once more considerable delay before PASOK was able to set up the legal and institutional components of the socialisation policy (see Mitropoulos, 1986, pp32-3). The eventual schema was the result of the work of a number of advisory groups working under Arsenis at the MNE, with the result, as we shall see, that there was little discussion of the policy within PASOK and that this schema was not universally accepted within PASOK as the 'correct' interpretation of the general principles of its socialisation policy.

Law 1365/83, which set out the basic principles and objectives, was not passed until 1983. The Law stated that for 15 DEKO there would be individual Presidential Decrees which would lay out the institutional arrangements for the socialisation of each DEKO. In fact the first three Decrees were only signed in October 1984, with the result that the various organs envisaged were not able to begin operating until after the 1985 election. The basic institutional arrangements for the first three DEKO (DEH, OTE and OSE) to be socialised were:[14]

1. <u>The Representative Assembly of Social Contract (ASKE)</u>: this body consisted of 27 representatives from the Greek State, the employees and various social bodies (eg representing consumer interests). This body was responsible for the medium and long-term planning of the enterprise, for invoicing of the services and for procurements policy. It was also charged with supervising the implementation of the enterprise plan and assessing whether plan targets had been achieved. It was envisaged that the ASKE would also have an input into the

preparation of the national five-year plan (for instance, the ASKE of DEH influencing the plan in the area of energy). There were also PESKE set up which were similar bodies to ASKE at the decentralised/regional level operations of the enterprise.

2. Board of Directors: this body consisted of nine members, six elected by the government, and three by the employees. It was charged with supervising the management of the enterprise without getting involved in every-day management decisions. For that purpose it elected a General Manager, who appointed a Management Council.

3. Labour Councils: these consisted of representatives from the employees and had a strictly advisory role on general matters of management, operation and control. They were intended to suggest means to improve productivity and the quality of services, as well as to advise on industrial relations.

A number of points can be made about this schema. PASOK ministers argued that socialisation entailed neither self-management nor nationalisation. The former was rejected because of the size of these firms and the importance of externalities, for instance in energy, telecommunications and transport, with the result their operation affected the wider economy (see chapter 3.1). The inclusion in ASKE of other interested bodies, such as local authorities, was to avoid some of the pitfalls of 'enterprise syndicalism' (see chapter 3.4; Simitis in MNE, 1987, p15-19).

However the fact that the state must have overall responsibility in these firms, did not mean that this power could not be socially controlled. The importance of the ASKE lies in the fact that it represented PASOK's response to some of the problems, discussed in chapter three, with respect to the control of public enterprises.

Firstly, the ASKE could help to promote the autonomy of the DEKO and transparency in their operation, by supporting the government's task of supervision. Secondly, the ASKE could aid the government's strategy of promoting enterprise planning within the DEKO. We saw in chapter three that such goals are not in fact easily realisable - the ability to exert political pressure and appoint managements gives the state limited power in confronting the bureaucracy of firms which have their own priorities, and are able to obstruct government policies. PASOK's response was that the ASKE could exert pressure on the management to establish institutions for enterprise planning and supervise its progress.

Thirdly, the ASKE could help in integrating the DEKO into planning. Papantoniou, a junior minister at the MNE after 1985, outlined PASOK's conception (MNE, 1987). The ASKE should provide an input into the five-year plan in their respective fields. The government, taking into account the views of the ASKE and the priorities of the national plan would formulate the general guidelines for each DEKO in terms of investment and prices policy. The ASKE would take these guidelines in order to provide a medium-term plan for the enterprise, instructing the Board to present a programme for the implementation of these goals, as well as for the organisational restructuring and enterprise planning of the firm. Finally the ASKE would have to agree that these programmes were indeed consistent with the medium-term plan, and would then supervise the implementation stage.

Fourthly the ASKE schema, by rejecting the previously authoritarian structure of these firms, was intended to mobilise the responsibility and dynamism of workers, thereby contributing to economic efficiency.

These four elements therefore represent a relatively sophisticated response to the problems of control, and the integration of state firms into planning. If PASOK rejected the self-management model, it also rejected the Morrisonian model of nationalisation. The ASKE could provide a form of social control that went far beyond the rather formal control of responsibility to parliament through the relevant minister that characterises Morrisonian nationalisation.

6.3.3 The implementation of the socialisation policy

The socialisation schema was subject to a number of problems during its implementation stage which prevented its effective operation. We describe here the most significant of these problems before going on to draw some conclusions of more general relevance.[15]

Before 1986 there was a lack of clearly formulated government plans, within the framework of the 1983-87 plan, for the operation and development of the DEKO. This severely limited the contribution that the ASKE could make, preventing any integration of the DEKO into the national plan or any serious reorganisational restructuring of the DEKO themselves (see Papantoniou in MNE, 1987, p21). There was some reversal of this in PASOK's second term with the adoption in 1986 of the MNE's 'reconstruction programme', to be discussed below.

The second problem relates to the delays in providing some of the administrative prerequisites for the operation of the ASKE. Papanikas, President of the ASKE of DEH, pointed to a number of examples: the limited amount of help given by the government to the ASKE in terms of providing resources, information and lines of communication; the reliance on ad hoc meetings and exchange of views; the reluctance of the Boards to integrate the ASKE into all levels of the enterprises' operation and their failure to provide the ASKE with reports, information and guidelines; and the hostility shown by many managers in the enterprises to the role of the Labour Councils (see Papanikas in MNE, 1987, pp31-35). Representatives of ASKE were quick to point out that the ingrained bureaucratic practices and interests of the DEKO were bound to find the ASKE 'troublesome', but that it was in the very nature of the ASKE to be 'troublesome' if they were to promote organisational restructuring and the transformation of old practices. Furthermore if the ASKE were to go beyond their role of 'shaking up' management and the organisational framework of the DEKO, they would need the type of technocratic support facilities, including research personnel, that were lacking.

The third related problem was the rivalry of various organs, within the socialisation schema, attempting to expand their sphere of responsibility at the expense of other organs. For instance, there was a tendency for the Board to intervene in the daily management decisions which the socialisation process was intended to prevent. The government was eventually to attempt to reform the existing Presidential Decrees, to limit the power of the President of the Board and enhance that of the General manager.[16]

However the most severe conflict concerning spheres of responsibility came over the role of the ASKE. The occasion of this conflict arose when the government, within the context of its October 1985 stabilisation programme, attempted to enforce a financial reconstruction of the DEKOs debts by pushing through, amongst other measures, price increases (albeit at rates lower than inflation) in the services/utilities of the DEKO. Law 1365/83, and the Presidential Decrees, stated that the ASKE had exclusive responsibility in medium-term planning with respect to investments, price policy, etc. At first the government attempted to push these price adjustments through the ASKE in order to promote consensus over the policy. However the ASKE complained that often the government had not even consulted them on the price rises or presented any detailed documents outlining the rationale behind them (see, for instance, Batistakos in MNE, 1987, pp83-86). Even worse, the government at times, by manipulating its representative on the ASKE, was able to prevent a quorum at ASKE meetings, thus preventing any debate. The subsequent conflict between the ASKE and the government, led the latter to attempt to amend the Presidential Decrees to limit the role of the ASKE. In the event the amendments were not pursued by PASOK partly because they encountered constitutional difficulties, but more because the dynamism of the ASKE had stagnated.

6.3.4 Socialisation, labour participation and the ASKE

The above problems in the operation of the socialisation policy severely affected PASOK's ability to integrate state firms into the planning process. By analysing this phenomenon, we can reach some conclusions of more general relevance.

In an important article, Murray (1987) has discussed a number of formulations that have been adopted to integrate nationalised industries into planning and overall government policy. He asks the question of how a progressive government can ensure that the state managers put its policies into practice and discusses three dominant approaches: the distinction between strategy and implementation, the tradition of a neutral administration and the development of methods of accountability which operate as if they were the market. He goes on to examine the inadequacy of all three (Murray, 1987, pp108-112). He argues that the reliance on a neutral administration is difficult for a socialist government in that it would confront the pre-existing class relations within the public administration (eg salary levels, qualifications, relationships with private sector interests). Methods which impose controls on administrators to mimic market controls, such as financial targets or performance criteria, are also subject to numerous problems. These include the need to integrate economic and social goals, the difficulty in getting agreed criteria on such issues as performance, and the ability of firms, with their control of information, to dictate criteria or norms best suited to themselves (Murray, 1987, p100). Finally Murray is critical of the relevance of the distinction between strategy and implementation: 'economic planning is not like this. It is a question of guessing what is over the horizon, of adjusting strategy in the light of practice, of assessing the political as well as the economic possibilities of further advance' (Murray, 1987, p97). Murray thus rejects this traditional method for integrating state firms into planning in favour of a conception which rests on a much greater interplay between strategy and practice, entailing resources for workers, consumers and other interested parties to put an input into planning.

Such a conception gives us a useful framework in which to analyse PASOK's socialisation policy. While we have been discussing the socialisation schema being promoted by the MNE, there was no uniformity of opinion within PASOK on the appropriate policy for the DEKO. Thus side-by-side with the MNE approach, we can delineate three other approaches. One view envisaged a rather moderate role for the ASKE, exercising the type of control which would mimic that of shareholders' meetings. Diametrically opposed to this view was that held by those who sought to 'deepen' the socialisation process in the direction of self-management. Yet a third view was held by those PASOK ministers, or administrators, who opposed the socialisation policy for the reason that it restricted their room for political intervention - in the appointment of personnel, in intervening in DEKO management, in promoting social and electorally-orientated policies.

Moreover, there was no process, or forum, within PASOK where such issues could be openly discussed and a compromise reached. Various groups within PASOK tried to further their conception in areas where they had some influence. Thus pro-PASOK members on the Board of a socialised DEKO might have a very different conception to that of the PASOK ASKE representatives, contributing to the political conflict we have already described. This, and the lack of debate, made it impossible for PASOK to set out the ideology of the socialisation policy, to attain a degree of support for the policy on which its success crucially depended. In turn this could have led to an attenuation of the conflict between the various organs within the socialisation schema.

The policy of socialisation could not rely on a neutral

administration for the reason that in Greece, both within the government administration and the DEKO themselves, there existed an extreme lack of modern organisational methods, the absence of clear channels of responsibility, and a tradition of using the public sector as part of the social and employment policies of successive governments (see Tsoukalas, 1986). PASOK actually exacerbated this phenomenon for, as Arsenis has said, the most significant mistake of PASOK's policy for the DEKO was the continuation of the practice of appointing personnel on party political grounds (Arsenis, 1987, pp103-9). While such problems may be particularly acute in the Greek case, there are good reasons to think that similar problems would be confronted by any left social democratic strategy elsewhere. The critical weakness in the model of planning, discussed in chapter three, is the lack of a theory of public administration. Murray's rather enigmatic comments on the need for creating socialist planners, while no doubt an important aspect of this problem, cannot hope to cover the issues involved in creating a public administration suitable for the type of planning involved in alternative economic strategies. The lack of a theory on such issues was a fundamental flaw in PASOK's strategy, but it is likely that this would also be the case in other countries as well. We return to this issue in the concluding chapter.

Of course an essential element of PASOK's response to such problems was the socialisation policy itself, and in particular the role of ASKE in carrying out an element of social control. What can be learnt from the experience of the operation of the ASKE? Let us return to the conflict between the ASKE and the government over the question of the price increases in DEKO utilities/services. The MNE position was that the ASKE could, and should, have only an advisory role on short-term economic policy matters. Given their size the ASKE are not appropriate bodies to intervene in daily management decision (Papantoniou in MNE, 1987, p23). There are indeed some sound economic reasons why the ASKE should only be involved in medium-term planning. Consider the pricing policy of a particular DEKO. Any medium-term plan would of course have to deal with this matter. But the management will also have to respond to short-term economic conditions. Thus a firm facing higher interest rates in a particular year, may legitimately seek relatively higher price increases in that year, and relatively smaller ones in the next, while keeping to the overall price policy of the medium-term plan. In short, medium-term planning has to do with the trend of the system, while short-term economic decision-making has to do with contingency, with the cycle. The latter entails a degree of flexibility which a body similar to that of the ASKE cannot hope to achieve. In drawing this distinction, the PASOK government was essentially correct.

It will be noted that this distinction between medium-term planning and short-term economic decision-making is not the same as that between strategy and implementation, previously discussed. For the ASKE could provide a forum for the type of enterprise planning, with an interplay between strategy and practice, which Murray has supported. Only this would have to be with respect to medium-term planning, and not management economic decision-making. Whatever the appropriate distinction that needs to be made there is a more general conclusion. Many models of alternative planning emphasise the importance of social control and workers participation (see chapter 3.4). The experience of the ASKE, however, shows the need for clarity and precision in the institutional framework to promote such planning, in which not only the ideology and goals of the planning process are clearly elucidated, but the role of the various new institutional organs are specified and their relationship to each other clearly indicated.

While we would argue that the ASKE formula did provide a basis for an increased element of social control and worker's participation into the DEKO and planning there were a number of other considerations why in the Greek case such a conception could not be furthered. By examining these we can perhaps shed light on some prerequisites, of general validity, for such a project.

The first aspect was that under PASOK, the political climate was not favourable for the active involvement of workers. This phenomenon began with the inclusion in Law 1365/83, of Article four which severely limited the right to strike in socialised industries. The pill was made harder to swallow by the fact that Article four's operation began immediately, while the other aspects of the Law were not introduced until after the 1985 election. Furthermore, the socialisation schema also created an element of uncertainty in that all labour representatives (in the ASKE, Boards, and Labour Councils) were elected directly and were not subject to accountability procedures to their base. Indeed PASOK seems to have explicitly intended to by-pass the trade union movement (see Papantoniou in MNE, 1987, p20). This dualism was bound to create problems (Mitropoulos, 1985, p125). The socialisation schema was thus not conceptualised as an extension of collective bargaining, to such questions as investment and planning.

In the post-1985 period, the relationship of PASOK and the trade unions, with a strong communist and PASOK defectors presence, deteriorated seriously. This was the result of the adoption of the stabilisation programme and what the trade unions saw as PASOK's abandonment of socialist interventionist policies. After the election PASOK lost its majority on the GSEE (the Confederation of Greek Workers), which was reimposed only after judicial intervention. Whatever the possible responsibility of the government in this decision, the least one can say is that PASOK, as a political party did not attempt to resolve the succeeding crisis. The subsequent loss of legitimacy of the GSEE in the following years hardly provided the context for serious worker participation in the socialisation process.

However not all the responsibility for this can be laid at PASOK's door. The type of workers participation involved in Murray's conception of planning entails a form of trade unionism which was simply lacking in Greece. We have previously referred to the weaknesses of civil society in Greece and the lack of horizontal linkages within the social formation. A long history of illegal, or semi-legal, existence, together with the policy of successive governments in controlling trade unions from the centre, has clearly left a large imprint on the Greek trade union movement. The struggle to gain legitimacy and defend minimum standards of living create the type of ideology and organisation which is not necessarily ideally suited to trade union intervention in the planning process. Naturally many in PASOK were aware of this from the beginning. As Arsenis argued one could not wait until the existence of a 'perfect' trade union movement, but had to promote those institutional mechanisms which would provide the framework for the changing role of trade unions.

However it is important to note the scale of the problem. In an article on the Labour party's planning proposals of the late 1980s, Lane (1987) examines the potential of British Trade unions to intervene in such areas as the proposed tripartite sector working parties of NEDC (which Labour intended to turn into bargaining planning forums). He points out the degree to which the attempt to expand collective bargaining to issues of investment, training and development poses a new set of priorities and problems for trade unions. He goes on to discuss the organisational (organisational structure, resources available to collect and analyse economic indicators etc) and ideological prerequisites for such a role for trade unions: 'the underlying cause of union's organisational weaknesses is that they have not normally and as a matter of habit concerned themselves with the logic of the enterprise. Regardless of their politics, trade unionists have habitually responded to managerial initiatives and tacitly accepted enterprise planning as the legitimate sphere of management' (Lane, 1987). If this is the case in Britain, then the constraint imposed by such factors in the Greek case are bound to be far greater. And they are very real constraints, imposing severe limits on what PASOK could possibly be expected to do, at least, in the short term.

6.3.5 Last attempts at a DEKO policy, 1985-87

Given the contradictions in the socialisation policy and the lack of some of the crucial prerequisites for its success, it is not surprising that PASOK shifted the emphasis of its policy after 1986. While the previous approach was based on the dual rationale of social control and economic efficiency, the emphasis was now very much on the latter.

The government set up a General Secretariat of DEKO[17], at the MNE, to provide a framework for rationalising government supervision of the DEKO and to work out an economic and organisational programme for their sectors. These programmes had three major goals: a) to ensure the DEKO managements implemented specific measures for cost reductions and improved services; b) to promote enterprise planning within the DEKO; and c) to carry out financial reconstruction through price adjustments and increased transparency in the amount of subsidy with corresponding transfers from the state budget. This package, aimed at restricting the gross borrowing needs for the largest eight DEKO to 2% of GNP by 1989 - as opposed to 7.2% on current trends - was adopted by the Cabinet in November 1986. The framework for government intervention was also clarified through a process of yearly agreements between the Secretariat and the DEKO on budgets, social budgets (to distinguish the social policy aspects of the DEKO from their main entrepreneurial activities) and the building up of the technical and administrative infrastructure for the Secretariat's supervision.

This new 'rehabilitation programme' and new schema for government supervision was implemented for about two years. According to the MNE press release on 4 August 1988, Public Corporations' deficits as a percentage of GDP were cut by two percentage points between 1985 and 1987. However there was much less success in respect to the deficit of the Social Insurance Organisation which continued to grow.

The relative success of the rehabilitation programme with respect to the Public Corporations reflected the strong modernising streak at the MNE, under Simitis and Papantoniou in the post-1985 period. However this period was short-lived and was terminated at the end of 1987 with the resignation of Simitis from the MNE.[18] But for our purposes it is important to point out the extent to which the ASKE had been by-passed in this process. The government argued that this was not necessarily the case - the ASKE could still have a central role in putting pressure on the managements to comply with the rehabilitation programmes and the adoption of corporate planning (Papantoniou in MNE, 1987, p23). However the rehabilitation programmes were never discussed with the ASKE, and these programmes covered issues on which the ASKE were supposed to have a 'decisive' influence - the 'rehabilitation plans' essentially became the medium-term plans of the DEKO. Thus although with respect to its policy to the DEKO, PASOK had a clearer conception of its strategy than it did with respect to either planning agreements or public procurements, the policy here also did not add up to very much in the final analysis. Indeed after 1987 interest in integrating the DEKO into planning, the ASKE and so on fizzled out completely.

6.4 Conclusions

In this chapter we have been examining the institutions and mechanisms of PASOK's alternative economic strategy. If PASOK had an overall conception of its strategy in 1981, this did not include a detailed operational plan which could be implemented relatively quickly and confront the various problems that could, and should, have been anticipated. Thus an adequate framework for planning agreements and socialisation was not set up until after the 1985 election, thereby severely limiting the effectiveness of the 1983-87 plan. It seems that PASOK had severely underestimated the scale of the problem involved in setting up the legal and administrative framework of planning - a task which had to be initiated almost from scratch. Furthermore, while some of these planning institutions/mechanisms were introduced in a more coherent form after 1985, there was by then considerable doubt about how high on the scale of PASOK priorities planning rated after the change of course with the stabilisation policy (October 1985).

The monumental task of setting up the administrative and legal framework of planning was not made any easier by the lack of co-ordination in PASOK's policy during its first term. One major problem was the lack of an institutional framework for PASOK to resolve its differences either at party or government level. We have seen this with respect to continuous ministerial interventions in the work of KEPE and the Nomoi Councils, in clientelistic politics in public sector recruitments and in the conflict around the progress of PASOK's policy of socialisation of the DEKO.

The lack of co-ordination prevented PASOK clarifying the central priorities of its strategy, within the framework of the plan. This severely limited the state's ability to carry out large industrial projects or to extend social ownership, within the framework of Law 1262/82. We have argued that the 1983-87 plan, by seeking to be comprehensive, in fact obscured the need to set out economic priorities and to consider the costs of the necessary process of restructuring the Greek economy. Had such priorities been delineated, planning might perhaps have been able to concentrate on those areas. This in turn may have provided a process of learning from the experience of that planning and may well have led to more support for planning as people came to see its successful implementation.

There are a number of other conclusions of more general relevance here. The type of decentralisation and less comprehensive planning, discussed in chapter three, depends on a number of prerequisites that were lacking in Greece. Thus the model of local enterprise boards and local planning crucially depends on a certain amount of autonomy for local authorities and planners to be able to promote a strategy for their own community to impart a strong dynamic to the local economy. The strategy of socialisation crucially depends on, not only the autonomy of trade unions, but their ability to restructure their ideology and organisation to enable them to provide a constructive contribution into enterprise planning. We have seen that the weakness of Greek civil society, severely limited the level of autonomy. While, no doubt, such problems are more acute in the Greek case, their importance to the overall coherence of the left social democratic model suggests that this would also be a severe obstacle elsewhere.

Certain other prerequisites for this model of planning have also been indicated. In this chapter we have stressed the inability of PASOK to reform the public administration, to make it more sensitive to the demands of planning. The lack of a theory on public administration and how it can relate to the needs of planning is therefore one of the most significant 'black boxes' in any interventionist planning strategy. The problems of building up a process of planning in Greece was related to problems of technical and administrative infrastructure, organisation and politics. We have suggested that all three were, to an extent, socially determined. Thus PASOK's failure to coordinate its planning policy and to clarify its central economic priorities had a further consequence. The national plan in Greece never became a focal point for economic decision-making of either the public, private or social sectors. This severely limited the popular support or consensus for planning, which is perhaps one of the most important prerequisites for successful planning.

Notes

1. The latter can be usefully compared with left social democracy's commitment to encouraging a wide range of social ownership forms (see chapter 3.1).

2. The crippling effect of this phenomenon is also discussed in chapter seven with respect to PASOK's policy for the 'problematic firms' and its sectoral industrial strategy.

3. One of the aims of the strategy of socialisation of the nationalised industries (see section 6.3) was to integrate nationalised industries into the planning process. However, as we shall see, this strategy had little relevance for at least the 1983-87 plan.

4. This comes out from nearly all interview conducted with KEPE officials. Ministers, it seems, were able to insist on the inclusion of certain local development projects to promote 'their patch'. Sometimes the government intervened to include a project for political or electoral purposes - the most obvious example was the sticking on of the proposals for the Athens metro literally at the last moment before the publication of the plan.

5. The divergence between these two figures is the result of two factors: firstly that some projects were planned to be implemented over a longer period than the five years of the plan; and secondly the plan accepted that for some projects the necessary technical studies had not yet been prepared and thus more projects were included than would probably be implemented.

6. Papagiannakis presents a synopsis of the history of these proposals and the numerous feasibility studies undertaken at various times.

7. It is probable that the new emphasis on the large infrastructural projects was related to a re-evaluation of the importance of services in the Greek economy, as expressed in Papandreou's much publicised Davos (in Switzerland) speech. For another example of such new

thinking, see Halikias (1988).

8. The main planning agreements in the first term concerned the fields of energy, transport and telecommunications. For a list of planning agreements actually implemented in the first term, see MNE, 1986, pp55-6.

9. From interviews conducted it seems that Vaitsos was fairly successful in setting up the necessary organisational and technical infrastructure and there does not seem to have been any obvious rational consideration, in respect to policy coordination, in this administrative reorganisation. The cause of this development has more to do with the sensitive nature of the public procurements budget and inter-PASOK rivalries at the political level.

10. See, however, the interesting article by Danos (To Bima, 29/3/87), who gives an account of the success of the 1986 public procurements programme and discusses the prospects of the 1987 programme and the new institutional framework.

11. For an extensive portrayal of the details of the new regulations of Law 1262/82, see Commercial Bank of Greece (1982c). For a more critical appraisal, see Hadjisocrates (1982).

12. Konstantinidis (1987) estimates that of the 32 billion Drachmas of endorsed investments by March 1987, 9 billion were for this 'new class of investors'.

13. See Kaloudis' 'Anti-incentives for Investments' in To Bima 2/9/84 for an extensive discussion on the private sector's position, and its opposition to state intervention and the macroeconomic and incomes policy of the government.

14. For an extensive treatment of the institutional arrangements of both the law and the Presidential Decrees, see Mitropoulos (1985; 1986).

15. This section is based on both interviews and a conference held at the MNE, with representatives from the state ASKE, Board of Directors and Labour Councils, to assess the course of socialisation one year after it had begun to

operate (see MNE, 1987). See also the special issues on socialisation in Sindicalistiki Epitheorisi, no 4, April 1985; no 22, October 1986.

16. This in turn was opposed by many labour representatives, since it would weaken the position of the Board, which has three labour representatives, and strengthen that of the management executive team which has no labour input -see Panagiotakis, in Demosios Tomeas, no 14, November 1986, pp46-9.

17. See article in Demosios Tomeas no 6, March 1987, for the structure and responsibilities of the General Secretariat. The first report of the Secretariat can be seen in Demosios Tomeas no 17, February 1987.

18. The resignation resulted from a dispute between Papandreou and Simitis on the extent to which the 1985-87 stabilisation measures should continue. Clearly Simitis was more committed to the stabilisation policy, with which he was strongly associated, than was Papandreou. Indeed Simitis' resignation, in retrospect, represents the beginning of the last period of PASOK's government, after which it is difficult to detect any attempt at a rational modernising economic policy, let alone a commitment to an alternative economic strategy.

7 Industrial strategy and the problematic firms

While the national plan set out the overall goals for industrial restructuring, the specifics of this was to be made clear by a set of complementary policies for intervention in industry and manufacturing. PASOK attempted to establish a framework for planning at the sectoral level and set up a state holding company not only to restructure certain 'ailing firms' but also to promote the government's overall industrial policy. Therefore here too PASOK's strategy can be seen as an important test case for such mechanisms and institutions envisaged by our model of planning in chapter three.

In section 7.1 we briefly outline some of the structural problems faced by industry in Greece, both at the level of the firm and the institutional framework within which firms operate.

In section 7.2 we examine two aspects of PASOK's approach to industry, supervisory councils and sectoral planning.

In section 7.3 we consider, at greater length, PASOK's response to the 'ailing industries' phenomenon and the setting up of the state holding company with the task of restructuring these firms and assisting PASOK's overall industrial strategy. The case study is employed to shed light on the potential and limitations of such holding companies as a component in alternative economic strategies.

7.1 The structural problems of Greek industry and the 'ailing industries' phenomenon

In chapter five we noted that by 1981 PASOK was facing a serious economic crisis, expressed in the deterioration of all the main macroeconomic indicators. In particular we pointed to the deterioration in the structure and pattern of growth and investment, and the extent to which industry was returning to traditional activities, specialising in products with slowly growing markets. By investigating the underlying causes of this phenomenon, we can examine the structural weaknesses of the Greek economy and the scale of the problem confronting PASOK's interventionist strategy to promote the supply-side response of the Greek economy. We focus on the emergence in the 1970s of a large number of 'ailing industries' - firms facing severe difficulties in repaying their debt to the financial sector. The underlying causes of this phenomenon can be analysed at various levels - at the level of the firms themselves and the institutional framework in which they operated. A full account of the structural weaknesses of Greek industry is beyond our scope here. Rather our task is a more modest one: to argue that Greek industrial/manufacturing firms faced a set of problems which the type of strategic planning, at the level of both enterprise and sectors (see chapter 3.1), was supposed to confront.

<u>Firm organisation</u> A crucial characteristic of the industrial structure in Greece by the late 1960s was the co-existence of a large number of small, often family-based, firms, with those larger firms, mostly in more capital intensive sectors (petrochemical, plastic and metal industries), originating from the Greek industrialisation of the 1960s and a new wave of foreign investments (see Giannitsis, 1986; Vaitsos, 1986). While this has often been characterised within the conceptual framework of the dependency school, more recent studies on the concentration of Greek industry have revealed that up to the mid-1970s, both small and medium-sized firms underwent significant capital accumulation, without the type of polarisation which would be predicted by a dependency-influenced analysis

(Doxiadis, 1984; Lyberaki, 1988).[1] Thus Doxiadis convincingly argues that the specificity of these Greek firms lies in their level of internal organisation and their articulation with the institutions of development.

There is a widespread consensus on the contribution of organisational underdevelopment to the 'ailing industries' phenomenon (see Arsenis, 1987, pp109-29; Papandropoulos, 1984). The family nature of these firms restricted the use of efficient managers as the result of the owners' desire to keep the management of the firm within the family. From our calculations of 126 'ailing firms' we found that a disproportionate number were made up of firms in the process of growing from the small to medium category and consequently facing more acute problems of organisation, marketing, industrial relations etc. Qualitative and empirical research has suggested that few of the 'ailing firms' undertook serious corporate or manpower planning, market research or technical-economic studies of proposed investments (see Tsotsoros, 1985; Papandropoulos, 1984).

Thus low Greek competitiveness is not only due to questions of high unit labour costs. Greek manufacturing products suffer from a lack of diversification towards more sophisticated and technologically advanced products (OECD, 1987, p30). In the worsening economic climate of the 1970s, Greek firms shifted to traditional activities, what they 'knew best', rather than undergo any serious restructuring and modernisation, or move into new products or sectors (see Papagiannakis, 1988). Firms lacked the organisational capacity to promote strategies for competing on non-price factors. It will be noted that such problems as product innovation and design, were stressed in those analyses, that emphasised the importance of the 'new competition' and 'flexible specialisation' (see chapter 3.1). It was this organisational failure which was seen as the rationale for enterprise planning.[2]

The institutional framework for development The organisational features of firms can only offer a partial explanation for the 'ailing industries' phenomenon. The voluntaristic nature of many of the critiques of the role of Greek capitalists, with the emphasis on their search for quick profits or failure to take risks, fails to pose the question of why such capitalists should rationally prefer short-term profits and ignores the institutional factors which allowed them to do so (see Petmetzidou-Tsoulouvi, 1984). Thus it is important to examine the institutional framework which determines the way agents within the economy react to both market forces and state intervention.

The responsibility of the banking system in the creation of 'ailing firms' is widely recognised. The governor of the National Bank, itself the most exposed institution in 'ailing industries' debt, pointed to a number of factors:

> the financing of overinvestment in sectors which have for a long time been seen to be facing structural difficulties, the covering of a large part of the capital for investments with short-term credits, the delay in completion of investment projects and the unjustified waste in resources during the construction phase led to the indebtedness of a large number of firms... the uncontrolled covering of short-term financial needs without an examination of the firms' stock policy, or an inspection of market and intermediary prices or even an elementary control of the credit policies towards their clients, contributed to the further indebtedness of firms. (Panagopoulos, quoted in Stergiou, 1985).

In part this can be explained by the intransparent, often personal relationship, of the banks with their clients, with the result that many decisions were based on non-financial criteria (see Stergiou, 1985). Furthermore monetary policy in this period relied on credit rationing and a complex set of regulations tying most of the banks' funds to public sector or other 'priority activities' (Halikias, 1987). The limited resources available tended to go to firms which had a privileged position with the banks (Doxiades, 1984). Other firms had to face liquidity problems or higher interest rates as a result of borrowing from 'unofficial' borrowing markets. 'Grey credit circuits' could take the form of using 'unofficial' lenders or round-tripping, by which firms with privileged access to funds, borrowed more than they needed in order to re-lend (OECD, 1986, pp57-9). It explains why firms in the 1970s could complain of high interest rates, when real interest rates were in fact negative, and how even a small increase in interest rates could have serious implications.

The organisation of most commercial banks was also problematic. They had developed neither the organisation nor expertise for serious project appraisal and supervision, relying instead on real estate and wealth as collateral. The short-term nature of the banks' calculations, often to cover firms' investment and debt repayment needs, resulted in a tendency of firms being unable to service their debt after a period of between two and five years (Papandropoulos, 1985).

As firms' financial situation deteriorated, the banks found themselves increasingly acquiring significant real estate holdings and having to take a role in the management of these firms. Given their organisational underdevelopment the banks were in no position to carry out the necessary restructuring of these firms, and carried out a short-term 'damage limitations' exercise by continuing to fund these firms. As a consequence the financial position of the banks themselves deteriorated rapidly in the late 1970s as they became increasingly overexposed and had fewer funds available for more efficient enterprises. By the late 1970s, the scale of the problem was beyond the capacity of the banks to intervene with a longer-term and more viable strategy.

Given the extent to which banks were state controlled in this period, it is clear that the 'ailing industries' phenomenon also reflects the lack of any coherent state industrial strategy. However this should not be confused with the question of state intervention in industry, which was not lacking in Greece, for 'while exercising protectionism, government statism has developed procedures for intervention and control in all sectors of economic life, thus piling up bureaucratic impediments in industry without ever pursuing a specific industrial policy ... industry was left to develop within a strange environment that combined free-market mechanisms, bureaucratic procedures, contradictions, incentives and favouritism' (Marinos-Kouris, 1985, p27). Such intervention and protectionism explains to a considerable extent why the banks allowed firms to invest in exhausted areas, or ones with poor prospects, and why firms were not encouraged either to move into new sectors, or at least to upgrade their organisation and products within a particular sector. Thus the privileged relationship between the banks and certain firms was often at the instigation, or encouragement of government and state officials. The organisational underdevelopment of banks also reflects the traditional use in Greece of the public sector, in this case the publicly controlled banks, as instruments of social policy. This could take the form of providing employment in the banking sector, mediated through clientelistic relationships, or making financial resources available for privileged interest groups, industrial sectors or even particular enterprises. That is, instead of seeing financial policy and the banks as institutions to further development, the state used them as instruments for income distribution (Doxiades, 1984; Papagiannakis, 1988). There was no serious attempt to respond to the serious structural weaknesses of Greek industry (Giannitsis, 1986; Dedousopoulos, 1987).[3]

It is only with the combination of the structure and organisation of the 'ailing firms' themselves and the institutional framework in which they operated that we can understand the nature of the 'ailing industries' phenomenon. This approach gives us the scale of the task confronting PASOK. A crucial test of its policy therefore would be not only how it approached the problem of the 'ailing firms' themselves but also how it intervened at the institutional level to initiate an industrial strategy to promote industrial development and restructuring. The left social democratic response, discussed in chapter three emphasised the importance of strategic planning at the level of enterprises, which lack the organisational capacity to compete on non-price factors (such as marketing, product innovation etc), at the level of industrial/manufacturing sectors and at the intersectoral level. Our account of the underlying causes of the 'ailing industries' phenomenon, at both the level of firm organisation and the lack of any coherent industrial strategy, suggests that the Greek case would provide considerable potential for the type of planning mechanisms and institutions discussed in chapter three.

7.2 Supervisory councils and sectoral planning

Here we examine two sets of policies which were supported by PASOK to initiate the type of extension of social control, intervention in industry and the harnessing of the private sector into planning discussed in chapter three. In particular, for our purposes, such policies were intended to transform the institutional framework in which firms operated to promote both social and economic goals.

7.2.1 Supervisory councils

In chapter 4.2.1 we saw that the supervisory councils were a 'development' of PASOK's original strategy of socialising the strategic sector of the economy. Papandreou declared that for the moment socialisation was to be restricted to the public sector, with the supervisory councils extending social control to the private sector. Nevertheless the importance of this new institution was emphasised: the law on supervisory councils was seen as on of the three most fundamental interventions at the institutional level (see Papandreou, 1983; pp36-7). It was to be the crucial mechanism for harnessing the private sector with the national plan, for specifying economic priorities and for rationalising state incentives to industry (see Anastasakos, 1985: Mitropoulos, 1985).

Law 1385/83 on the supervisory councils (SCs) was not passed until August 1983. It provided the general framework for the operation of the SCs which were to act as 'decentralised organs of social control'. It was made clear from the start that the SCs would operate 'outside' the firms, and that their role was primarily advisory and one of supervision. SCs would consist of representatives from the management and workers of supervised firms, the local authorities and the state. The details for particular SCs were to be given by Presidential Decrees which would specify the firms to be covered, the representation in the SCs and a more detailed account of their operation. In November 1983 the first, and as it turned out only, Presidential Decree was signed for the metal-mining sector in the Nomos of Euvoia.

A clearer perspective on the role of SCs can be gained by examining the goals of this SC. It was supposed to follow every aspect of the activities of the supervised firms, in order to propose measures for the harmonisation of the metal-mining sector with the priorities of the national plan. Furthermore the SC would: supervise the management of the mineral and other natural resources exploited by the supervised firms; study the prices of imported and exported goods and services; investigate the potential for coordinating and increasing the productivity of firms and recommend improved ways of exploiting mineral wealth; research ways to widen the markets for products; follow the financial needs of these firms and the use of credits; recommend improvements in industrial relations (Mitropoulos, 1985, pp71-2). Anastasakos (1985) claimed that these functions, taken as a package, could play a central role in PASOK's industrial strategy.

What can be said of the SC conception? Their role was strictly advisory, especially in providing the MNE with information and proposals, with no independent powers of their own to instruct firms or implement their own strategies. However even in their final version there was a clear scope for SCs. Their operation could have started a process of building up the necessary infrastructure for planning. Much valuable experience could have been acquired with their role of gathering information, providing an institutional framework for improving information flows and recommending policy packages.[4] For we noted in chapter three that one rationale for interventionist planning and the extension of social control was the market failure associated with the lack of information or the existence of asymmetric information.

If the SCs had been able to develop such a role, then one could envisage a strengthening of their influence. The SCs could have operated as a powerful instrument in PASOK's attempts to channel finance to priority areas. Financial assistance to particular firms may then have been conditional on the agreement of the SC, providing a form of 'commercial leverage' (see chapter 3.2.3). By investigating the potential of various sectors, the SC could have helped the government clarify its industrial strategy at the sectoral level.

However the operation of the first SC came to an abrupt halt before it had got under way. This led to a postponement of the operation of the SC Law (Mitropoulos, 1985, pp77-9). There was no further action taken in re-establishing SCs either before or after the 1985 election, even though PASOK was still committed to this strategy, as late as 1987 (see Simitis, 1987).

7.2.2 Sectoral planning

PASOK's industrial strategy during its first term was the responsibility of Vaitsos, who as an academic had been an advocate of interventionist planning. Vaitsos set up a number of working teams to investigate the structural weaknesses of various industrial branches and to recommend policy packages for intervention and restructuring (see Vaitsos, 1986; Marinos-Kouris, 1985).[5] By September 1984, six branch studies were published for the paper, textiles, steel, industrial ores, agricultural machinery and shoe/tanning manufacturing industries. It was also announced that the publications of other branch studies would follow (see <u>Athens Express</u>, 20/9/84).

Marinos-Kouris (1985, p28) mentions a number of goals for these branch studies: firstly, to disaggregate the general aims of the national plan by branch; secondly, branch policy would provide the framework for 'active' planning since 'it guides the activities and interventions of public agencies; it draws the guidelines on how the function of private enterprises will be in harmony with the general development objectives of the plan and will not be determined by 'free-market rules"; thirdly, it would contribute to planners ability to prioritise policies or branches, while mobilising resources to further industrial development, the competitiveness and vertical integration of Greek industry; fourthly the studies could point to the potential for investment in new products or technologies, or the technological or quality upgrading of existing products. It was further proposed that joint ventures, for instance to enhance exports, could be set up to promote intersectoral restructuring.

The above can be made clearer by examining some of the proposals of the branch studies themselves. The textiles and clothing branch study envisaged three instruments to promote restructuring, diversification, mutual specialisation and increased co-operation for firms operating in this branch: planning agreements, the support of various state organisations such as the Greek Export Company, and financial incentives. Further measures included the development of units for the production of equipment needed by cotton and spinning mills and weaving shops, and the establishment of a centre for experimental production of new products. Measures were also proposed for improving the institutional framework: establishing a Textile Products Design Institute and documentation/ information centres, marketing centres etc. Significantly the branch study accepted the overproduction in textiles/clothing, necessitating the need to shed labour, and proposed measures for labour retraining. Finally the study recommended the creation of a central administrative unit with responsibility for the implementation of the strategy under the general supervision of the MNE.

The above account gives an adequate picture of the policy envisaged and we need not expand on it. For the fate of the branch plans mirrors that of the SCs. The six studies remained 'on the shelves' up to the 1985 election and beyond, and the further studies were never published. After the 1985 election,

V Papandreou attempted to restart the process of creating branch studies, but once more with little success. The branch studies continued to be referred to by PASOK ministers, and were even used, in an ad hoc manner, to aid certain investment decisions. What the branch studies never became was a basis for a coordinated interventionist experiment in 'active' planning.

7.2.3 Assessment of the sectoral industrial strategy

How can we account for the failure of PASOK to carry out its policy for SCs and sectoral industrial planning? It will be noted that PASOK's approach resembles that of strategic planning. As a rationale for intervention it accepts the existence of market failure and organisational underdevelopment. By examining the features of specific branches it makes the clear need for an industrial strategy to accompany macroeconomic policy, for particular branches of industry face specific problems. Finally the approach could have provided a basis for rationalising state incentives and altering the institutional framework facing firms. Nor did this entail ignoring the market; indeed the strategy was geared to improving the long-run competitiveness of Greek industry. Thus by examining PASOK's failure in this area we can shed light on some of the necessary prerequisites for strategic planning.

The first issue concerns PASOK's ability to co-ordinate an industrial strategy. The most striking example of this is that the sectoral industrial strategy was not the only conception within PASOK of how to specify plan goals at a more disaggregate level. KEPE, under the leadership of Katseli, was simultaneously promoting a strategy of 'Integrated Network of Activities' (OSD). The idea rejected sectoral planning in favour of initiating project planning, by setting up a 'core investment' around which a network of complementary activities, such as research and development, finance, transportation etc, could be developed. The OSD approach, influenced by East European experience, also led to a number of studies from KEPE, although they fared no better than the branch studies, being used only occasionally to guide ad hoc investment decisions.

There was, therefore, a great deal of friction between KEPE and the MNE, but there was no ministerial forum in order to resolve the conflict to prioritise one strategy or reach a compromise. Given this phenomenon, there is a severe doubt over the extent that either approach can be said to represent the policy of the PASOK government as a whole. Thus the branch studies eventually came to be seen as the 'pet' project of Vaitsos - when Vaitsos left the government, it was natural to consider that the branch studies left with him.

However this development cannot be put down merely to a failure of political will. Once more PASOK seems to have severely underestimated the scale of the task, both in terms of legislation and administrative infrastructure. The SC legislation was not only delayed but met with considerable opposition from firms. The branch studies strategy clearly depended on the creation of an appropriate organisational and administrative framework. The working teams, starting almost from scratch, had been able to specify policy proposals and begin to tackle the problems at the inter-branch level to articulate an overall industrial strategy - not a minor achievement for two years work. But given the speed with which economic conditions change, there was a pressing need to set up the administrative capacity to be able to plan on a continual basis and bring branch plans up to date. At the most rudimentary level, this would require resources for staff and research capacity, which however, PASOK did not manage to promote.

This lack of co-ordination at the government level was paralleled by a similar inability to promote the necessary support and consensus for the industrial strategy within society. The opposition from the private sector to SCs may seem to make this an insurmountable obstacle. However the delays of PASOK in implementing the various components of its planning alternative, as well as the considerable uncertainty over the details of this policy, were hardly conducive to a satisfactory relationship with the private sector, let alone harnessing it to the priorities of the plan. Papandreou had frequently underlined the importance of planning providing a stable framework for both public and private sectors and clarifying the 'rules of the game' within the framework of the mixed economy (see Papandreou, 1984, p38-41). In this task, during its first term, PASOK was conspicuously unsuccessful.

The opposition to the OSD and branch studies approach was far less acute. The branch studies working teams did initiate discussions with private sector firms, but there was no institutional framework for this. Moreover, these discussions did not extend to consideration of how to implement the studies after their completion. This severely restricted the usefulness of sectoral planning since 'the problem is more political and institutional, and is confronted with procedures for consensus and bargaining among the interested parties, and much less a technical [issue] which is solved by <u>ex ante</u> planning' (Doxiades, 1984, p48). Not all this was PASOK's responsibility and we should add here the existence of the set of obstacles to such consensus-building which operate in the Greek case, discussed in the previous chapter.

However PASOK is responsible in the sense that it failed to take a stronger initiative. The working parties had argued in favour of an explicit political representation on the working parties to ensure some element of co-ordination with PASOK's overall political strategy and to enhance the political prestige of the branch studies. This was refused. How do we explain such reticence? We would argue that there were two central considerations. A higher level of political representation would only have made sense if PASOK, or the government, had clearly articulated priorities. We argued in chapter 6.1.3 that the national plan itself had failed to delineate such priorities. PASOK was unable to delineate such priorities, not least because at the level of intersectoral restructuring this would entail both political costs and resources for retraining, new investments etc. This theme is developed in the following section when we discuss PASOK's strategy for the problematic firms.

However the question of resources points to a more severe obstacle to PASOK's overall industrial strategy. For this strategy clearly entailed resources for new investment, sectoral technology enterprises etc. The failure of PASOK's macroeconomic policy, in particular to restructure public expenditure towards investment and control deficits, was bound to lead to very limited resources for industrial strategy. This together with the legislative and administrative obstacles encountered in setting up the new institutions for sectoral planning, entailed that PASOK severely overestimated the potential supply-side response of the economy, in particular over the short-medium term, on which its whole strategy crucially depended. This point is of general relevance - PASOK's failure to integrate its macroeconomic policy with its proposed industrial strategy points to the fact that such an integration is a vital prerequisite for any planning alternative.

7.3 Industrial intervention and the problematic firms

Law 1386 on the 'Business Reconstruction Organisation' (OAE) was seen by PASOK as the third fundamental intervention at the institutional level - together with the supervisory councils and the socialisation of the DEKO. The immediate goal of Law 1386 was to confront the problem of the 'ailing industries' phenomenon, discussed in section 7.1. It sought to clarify the process by which non-viable firms would be closed down while viable firms would pass into state ownership or control as part of a 'dynamic alternative solution for the rehabilitation of industry and the financial system' (Papandreou, 1983c, p38). The strategy was much broader in its scope than the rehabilitation of a few enterprises, encompassing a number of further issues: the role of state intervention and industrial strategy, economic planning and workers participation, structural change and development.

This makes PASOK's strategy a significant test case for our overall examination of left social democratic attempts to go 'beyond indicative planning' and in particular the capacity of state holding companies to intervene at both the enterprise and sectoral level, in order to carry out restructuring, on the lines discussed in chapter 3.2. Furthermore, PASOK's strategy incorporates the dual rationale approach of integrating social and economic goals: industrial restructuring was to be partially attained by extending social control in industry and promoting labour participation in the operation of restructured firms.

By examining PASOK's strategy and the difficulties it encountered, we can shed light on some important prerequisites for the successful operation of such an approach, as well as pointing to the unresolved problems of the left social democratic model of intervention as a whole.

7.3.1 Law 1386/83

By January 1982 the government announced a series of short-term measures to deal with the financial requirements of the ailing firms and that it was examining the institutional arrangements necessary for a long-term solution. As in other areas of policy, PASOK may have had an overall strategy by 1981, but this did not extend to an operational plan which could be implemented after its election victory. Thus Law 1386/83 was not brought before parliament until June 1983, with the result that the first articles of the law did not come into operation until December 1983, and the first board of directors of the new state holding company were not appointed until well into 1984. Not surprisingly none of the firms covered by the legislation had been financially restructured by the election of 1985. Before examining this phenomenon, we should briefly outline the main elements of Law 1386/83.

Article 2 stipulated that the OAE would contribute to the 'social and economic' development of the country by rehabilitating firms, applying foreign technology and developing domestic technology, and setting up new social or mixed enterprises. Proponents of the policy pointed to the relevance of European experience of state intervention in firms' rehabilitation, and an industrial strategy which employed state (or joint) enterprises.

Under the legislation, shareholders, trade unions, major creditors of 'ailing firms', and bodies such as the Greek Unemployment Fund, could apply for a firm to come under the auspices of the Law. Two conditions for entry were stipulated: the existence of financial difficulties and that firms were potentially viable in the long-run. Applications were assessed by the Secretariat of Problematic Firms, at the MNE. If firms were deemed non-viable, then they would be covered by the liquidation clauses of Law 1386/83. If firms were potentially viable, they would come under OAE control, now being referred to as 'Problematic Firms' (hereafter PFs). By May 1984, the Secretariat had investigated 116 firms, with eventually 43 firms coming under the auspices of Law 1386/83. The new state holding company, OAE, could now freeze principal and interest payment of these firms for a period of up to 36 months, although interest payments continued to be charged to their accounts.

The rehabilitation process envisaged that the OAE would appoint a new management for PFs, including one OAE representative, and one from the workforce. The task of the new management was twofold. Firstly, to introduce a series of short-term measures for technological and organisational restructuring to improve the operation of PFs, with the OAE providing financial and other forms of assistance. Secondly, more long-term plans for restructuring were to be worked out to propose to the new owners, once the restructuring of the debt of these firms had been resolved.

The financial restructuring involved compulsory increases in capital by existing shareholders, the conversion of part of the debt into equity, and usually, the rescheduling of the remaining debt (OECD, 1987, p35). The process involved the negotiation and agreement of OAE, shareholders and creditors, with a failure to reach agreement leading to liquidation. The creditors faced a choice of being paid 25% of their claim by the state, or converting their claims into equal value equity stakes. In practice, the banks converted about half of their claim into equity, receiving for the other half government guaranteed OAE bonds carrying the going interest rate. This entailed a sharing of the cost of adjustment between the state and the banks, with the banks being helped directly from the state budget, since many of their assets were unlikely to have sufficient returns in the short run. What was left vague in all this was the percentage of the old debt to be repaid by the future operation of the PFs.

What would be the status of these firms after their financial restructuring? Here the situation was less clear. A distinction was made between OAE firms and the socialised DEKO, for its was made clear that the former had to be active commercial concerns and operate within a competitive market framework (Arsenis, 1985b; Tsotsoros, 1985, p23). Thus, for instance, after financial restructuring such privileges as freezing of debts would be removed. The planning envisaged here is similar to the conception of strategic planning, discussed in chapter 3.1 - not the replacement of market forces, but intervention to promote the organisational capacity of firms' ability to respond to market forces and increase their dynamic competitiveness.

Since the state had borne the majority of the burden of readjustment the original conception was that the majority of the shares of the PFs were to be owned by the public sector. However it was repeatedly stressed that the future status of these firms would vary according to circumstance. Some would be socialised on grounds of social efficiency, especially those in the vaguely defined 'strategic sectors'. The fate of the rest would depend on the role they could play within PASOK's overall industrial strategy entailing in some cases a transfer of shares to workers' cooperatives, local authorities or even small private investors through the capital market.

This left the exact role of the OAE in the future rather vague, although it was envisaged that it would contribute to PASOK's overall industrial strategy, through its control of some of the restructured PFs and its ability to set up new enterprises or joint ventures in areas where the private sector was, for whatever reason, unable or unwilling to invest.

7.3.2 The co-ordination of the OAE strategy

In this, and the following section, we are concerned with the implementation of the OAE strategy. The theme in this section develops earlier arguments of PASOK's inability to implement its strategy in a coordinated manner. More specifically we present a set of arguments to explain this lack of co-ordination and focus on PASOK's inability to integrate the short and long-term aspects of its policy.

The most indicative aspect of PASOK's lack of co-ordination was it inability to reach final decisions on the status of firms after their incorporation into Law 1386/83. In a parliamentary speech in 1986 on the PFs, V Papandreou (1986a), the new industry minister was critical of the delays of the previous administration in which Arsenis was the central figure in control of economic policy. She particularly stressed the inadequacy of the viability studies and pointed out that even Arsenis now acknowledged that only half of the 44 firms were viable in the long run. Only one firm was treated under Article 7, Paragraph 3, of Law 1386/83, which provided for a process of liquidation, the rest coming under Article 7, Paragraph 1, which outlined the process of rehabilitation. Effectively, V Papandreou was claiming that the viability studies would have to be redone.[6] However while V Papandreou promised a speedy resolution to the financial restructuring of viable firms, the process was dragged out for the whole of PASOK's second term of office as well.

Such delay resulted in the mounting costs of PFs debt. Table 7.1 provides a fair picture of the size of the problem by 1987. Although the increasing debt was, in part, the result of the interest payments that continued to be charged (if not paid) rather than a deterioration in the operation of these firms, this still entailed a considerable financial cost. This clearly limited the availability of resources for new firms and areas. The financial position of the commercial banks continued to be precarious, restricting their ability to lend to more profitable and dynamic firms (OECD, 1986, p57).

Table 7.1
Problematic Firms (Drachma billion)

Date	1981	1983	1984	1985	1986
Total Number of firms [1]	44	16	38	43	44
Total employed (000s) [2]	33.0	30.9	30.3	30.9	28.0
Net Sales	63.8	78.2	85.2	120.8	145.3
Gross profit margin	11.1	7.5	9.1	14.4	24.5
Operating expenses	7.6	10.2	11.0	14.4	15.7
Operating income	3.5	-2.7	-1.9	-	8.8
Interest, exchange loss etc	11.9	45.4	50.5	55.7	57.7
Net income	-8.3	-48.1	-52.4	-55.7	-48.9
Accumulated losses	18.6	79.0	131.4	187.1	236.0
Total liabilities	115.3	201.9	263.1	342.9	349.6
Total assets	116.2	146.9	157.7	181.9	219.1
of which: fixed assets	71.5	104.5	106.8	112.3	100.7
	\multicolumn{5}{c}{23 viable firms (end-86)}
Total employed (000s)	27.0	26.7	26.7	27.1	26.1
Net Sales	56.0	71.5	80.8	112.7	136.4
Gross profit margin	10.0	8.0	9.9	14.7	24.8
Operating expenses	6.2	8.8	9.9	12.6	13.6
Operating income	3.9	-0.8	-	2.1	11.2
Interest, exchange loss etc	9.4	30.9	37.1	45.2	44.8
Net income loss	-5.6	-31.7	-37.1	-43.1	-33.6
Accumulated losses	11.0	48.0	85.1	128.2	161.8
Total liabilities	85.3	147.8	195.1	258.3	250.7
Total assets	88.9	117.7	130.1	150.1	188.6
of which: fixed assets	57.2	86.8	89.7	94.8	83.1

Table 7.1 continued
Problematic Firms (drachma billion)

Date	1981	1983	1984	1985	1986
	\multicolumn{5}{l}{21 non-viable firms (end-86)}
Total employed (000s)	6.0	4.2	3.6	3.8	1.9
Net Sales	7.8	6.7	4.3	8.1	9.0
Gross profit margin	1.1	-0.5	-0.8	-0.3	-0.3
Operating expenses	1.4	1.4	1.1	1.8	2.1
Operating income	-0.3	-1.9	-1.9	-2.1	-2.4
Interest, exchange loss etc	2.4	14.5	13.4	10.5	12.9
Net income	-2.7	-16.4	-15.3	-12.6	-15.3
Accumulated losses	7.6	31.0	46.3	58.9	74.2
Total liabilities	30.0	54.2	68.0	84.7	98.9
Total assets	27.2	29.2	27.6	31.0	30.5
of which: fixed assets	14.3	17.7	17.1	17.5	17.6

[1] Before falling under LD 1386/83 all firms (ie 44 firms) had (Drachma billion):
Accumulated losses 109.9 Capital and Reserves 26.6
Total Bank Loans 158.9
Other Liabilities 71.6
Total Liabilities 230.5 Total Assets 146.3
[2] Total Employment all firms covered under LD1386/83.

Source: OECD, Table 11, 1987 (from data submitted by national authorities).

PASOK's strategy entailed, and this was frequently emphasised in policy pronouncements, clear cut decisions on which firms were potentially viable after a restructuring process. It was further argued that this also entailed, and here there was a greater reticence by PASOK, certain short-term costs in the transitional period. For in the short-run the closing down of non-viable firms was likely to lead to social costs as a result of lay-offs, even if these effects could have been mitigated by policies for redundancy and unemployment benefits. The benefits of the OAE strategy, as a result of OAE investments in new areas and firms, would take time to materialise. There was a clear reluctance to accept such short-term costs. The more PASOK delayed in taking critical decisions, the more acute the problem became. During its first term of government, PASOK could claim, with some justification, that certain closures of non-viable firms were the result of the economic policies of previous governments. Furthermore the support for the OAE policy could have been enhanced if it had been accompanied with the restructuring of other firms or the setting up of new investments. But by 1985, when PASOK was claiming that only half of the OAE firms were viable, it was seen to be promoting the closure of firms it had previously declared viable and thus could not transfer the political responsibility onto previous governments.

Two further considerations, already discussed, exacerbated PASOK's inability to integrate short and long-term aspects of its policy. One was PASOK's lack of a clear conception of its development priorities and industrial strategy. For, as we argued in chapter 3.2, the organisational and financial restructuring of firms by state holding companies rests considerably on situating a particular firm within the needs and potential of particular sectors. The failure of the strategy of supervisory councils and sectoral planning was bound to restrict the effectiveness of the OAE. Secondly, the lack of such clear priorities allowed PASOK to continue to finance the mounting debts of the PFs and use the PFs as instruments of its social policy, for instance, to ensure employment opportunities. This phenomenon was self-reinforcing - the longer PASOK procrastinated, the more financial costs soared, and resources for other purposes were limited. The OAE could hardly progress into the other areas envisaged by Law 1386/83, when it was overloaded with the pressing demands of keeping so many firms in operation, some of which it was eventually admitted were non-viable.

However there is another consequence of PASOK's failure to harmonise the short and the long term and its inability to implement the strategy in a coordinated manner. It severely restricted PASOK's ability to overcome the opposition to the OAE policy. Here we briefly examine the opposition from both the commercial banks and the private sector (Arsenis, 1986; 1987, pp109-37). The commercial banks, even though largely state-owned, were reluctant to provide the necessary funds for the rehabilitation process. It is for this reason that, Arsenis claims, the MNE had to rely in April 1984 on a consortium of banks to provide the necessary funds, since the private, and smaller, banks were more willing to provide finance for PFs. For Arsenis, such reluctance explains the considerable delays in the rehabilitation process.

The second issue of conflict with the banks related to the capitalisation of the debt of the PFs. The issue here is complicated by the private nature of the negotiations between the MNE and the National Bank of Greece, and the confidential nature of the banks' assessment of the financial position of particular PFs. In the 1986 Parliamentary debate, already alluded to, Arsenis claims that in early 1985, the Governor of the National Bank rejected the proposals for capitalisation (Arsenis, 1986, p1546). Arsenis' proposals were in fact close to the schema implemented after the 1985 election which distributed the burden between the banks and the state budget. However Arsenis opposed a general overall treatment of capitalisation of debt, preferring to treat each PF separately. By this he hoped to make more transparent the 'closed system of power' which had led to a particular firm becoming 'problematic' in the first place, to clarify how much of the burden should be borne by the banks in each case and to put pressure on the banks to reform themselves. It was this approach that Panagopoulos, Governor of the National Bank opposed vehemently.[7]

However, the issue is more complex. For Arsenis claims that the National Bank of Greece effectively represented certain private sector interests and never fully articulated in public its real policy of minimising both the extent of the public sector intervention and the degree of burden to be borne by the banks (Arsenis, 1987, especially pp125-6). On the other hand V Papandreou, in her reply to Arsenis in the parliamentary debate, points to certain letters of the National Bank during 1983 and 1984 which seem to accept that it was willing to undertake some of the burden of the capitalisation process. She concludes that if the process had been undertaken then it would have been at a considerably smaller cost than the one facing the economy in 1986 (V Papandreou, 1986, p1548-9). Thus V Papandreou attributes a major share of the delay to the MNE itself. Given the confidential nature of much of the documentation it is difficult to adjudicate on this, although there is no doubt on the existence of both political opposition and MNE delays.

The hostility of the private sector to PASOK's approach was evident from the beginning. This was led by the Confederation of Greek Industrialists (SEB). SEB's position was stated clearly during the passage of the OAE law in Parliament in 1983, and was to remain remarkably consistent. It saw the approach as one leading to expropriation and socialisation. It proposed instead that each firm should be dealt with individually through the banking system and on banking criteria, and without a new bureaucratic institution (see Stergiou, 1985).[8] Much of the argument of the private sector revolved around the question of the unfair competition as a result of the privileges of the PFs (freezing of debt, government subsidies, etc).[9]

In fact, PASOK was remarkably sensitive to this issue. It responded by pointing out that the closure of non-viable firms would assist other firms within the same sector. Furthermore it could claim that OAE only took responsibility for firms which were 'problematic' in the first place, and that their privileges would be removed after the rehabilitation process. Tsotsoros (1985, p36) discusses a number of policies promoted by OAE to prevent 'unfair' competition including the prompt payments of firms' national insurance payments and the open tendering for the primary resources and goods needs of the firms.

The private sector opposition took the form of the old owners taking their case to the Council of State, claiming that Law 1386/83 was unconstitutional. This appeal was rejected on the grounds that the extent of the indebtedness of the old owners hardly provided a case of 'state expropriation' of their property (see Mitropoulos, 1985, p87-9). However this process undoubtedly added to the uncertainty of the OAE process. Later some former owners took their case to the European court and indeed by 1987 the government was examining reforms of Law 1386/83 to make it in line with European Community Law. Finally the old owners were able, in certain cases, to obstruct the rehabilitation process since Law 1386/83 was in fact conciliatory enough to allow them a say in that process.[10]

What can be said of this opposition from the commercial banks and the private sector? For Arsenis (1987, pp123-5), the problem lay in that both had independent access to the prime minister. He states as an example that while both he and Panagopoulos had access to Papandreou, the latter refused to come to a final decision: 'the problem was that in this case, as in others, there was no clear mechanism of resolving differences'. In 1985, Arsenis acknowledged that a precondition for the success of the strategy was an awareness of the existing opposition and the ability to implement it in a way which broke with establishment interests and refused to balance all sides in an attempt to please everybody (Arsenis, 1986, pp1545-6). However it seems that under Arsenis before 1985 such a strategy had not been fully worked out. It is perhaps significant that the OAE was given a banking function, implicitly recognising the inability of PASOK to control the largely state-owned commercial banks.

The result of PASOK's inability to co-ordinate the implementation of its policy was that two central questions, on which the viability of the whole OAE strategy depended, were never resolved. The first has to do with the ownership status of the PFs after their financial restructuring. The political controversy here, within the government, continued after the 1985 election with respect to the number of firms which would remain under OAE control.[11] This controversy climaxed in the spring of 1987, with the resignation of the OAE president, G Anomeritis, and in the summer of the same year when the new Minister for Industry, G Petsos, seemed to indicate that all 21 firms were open to private sector purchase.[12] Although this radical privatisation policy was partially reversed, perhaps due to the lack of a serious market for such firms in Greece, it added considerably to the atmosphere of confusion and instability under which the OAE operated.

The second question was the extent to which the OAE should have a major role in industrial strategy. Thus after the 1985 election, V Papandreou (1986b) indicated that PASOK was still committed to the OAE role as an interventionist institution of industrial policy. But this insistence was continually put in doubt not only because of the question of OAE ownership of PFs, but also the lack of any framework for OAE to set up new public enterprises and the general context of PASOK's overall economic policy after the stabilisation programme of October 1985.

Therefore we have seen that there existed a set of problems associated with PASOK's OAE strategy which further highlights our theme of PASOK's inability to integrate short and long-term aspects of its policy. In particular we have pointed to three crucial aspects: the failure to set out the priorities of its overall development and industrial strategy; the inability to accept the short-term costs of OAE policy; and the lack of a clear conception of how to overcome the opposition to the policy.

7.3.3 The Business Reconstruction Organisation (OAE)

The lack of co-ordination in PASOK's OAE strategy, the controversy over the status of PFs after rehabilitation and the role of the OAE in industrial strategy, had a further consequence. It prevented a sustained debate on the theory and practice of how the OAE itself should operate. Here we examine PASOK's strategy from the perspective of the OAE itself, the administrative and organisational problems associated with setting up such a state holding company, the relationship between the OAE and the PFs and the role envisaged for labour participation.

Our analysis here focuses on the arguments developed in chapter three concerning the distinction between ownership and control. We argued there that the ability of a left social democratic party to nationalise firms or establish interventionist institutions did not necessarily entail its capacity to control nationalised industries and state agencies. By looking at the way the OAE has functioned and the problems it confronted, we may draw some general conclusions of wider relevance for the problems associated with an interventionist strategy.

The first question of relevance here is the organisational problems of the OAE and its relationship with the MNE. We have seen that Law 1386/83 was seen as an institutional intervention to confront the existing bureaucratic and inflexible public sector. But of course the establishment of a new interventionist institution is associated with its own set of problems including finding the appropriate personnel and clarifying the framework for its operation. The legal problems and opposition confronted by Law 1386/83, is merely one side of this. The process was bound to be a long-term and intricate affair, relying on a growing body of experience. Finding the appropriate staff was particularly difficult, given the lack of experience, in the Greek case, of any similar experiment. Although a large number of economists and post-graduates from other academic fields were available and employed, there was a lack of people with experience in the practical aspects of production, market research, and management (see Nikolinakos, 1986). There was also difficulty in the appointment of OAE managers for the PFs, a problem that was increasingly stressed by PASOK in its second term (see V Papandreou, 1986b). In the first term this was exacerbated by the limits imposed on the OAE with respect to salaries, although this constraint was eventually relaxed (see OECD, 1986, pp35-6).

The OAE had from its inception a strictly limited autonomy in its operation vis-a-vis the MNE. The viability studies on ailing firms were undertaken by the Secretariat at the MNE, with the result that the OAE had no control on which firms come under its auspices. The government failed, beyond the general principles of the legislation, to articulate a consistent set of policy guidelines for the OAE. Its inability to reach final decisions on the status of PFs and the failure to promote other aspects of its industrial strategy, such as sectoral industrial planning, made the OAE's task almost impossibly difficult. In this context the OAE lacked the autonomy necessary to promote its own corporate loyalty in favour of planning and articulate an independent strategy for intervention within the framework of the legislation.

The above problems of control suggest, once more, that PASOK severely underestimated the legal, administrative and organisation preconditions for its overall strategy. However a further problem of control was confronted with respect to the relationship of the OAE with the PFs. In our discussion in chapter three we examined the strategic planning conception which entailed, given the importance of the 'new competition', responding to the organisational underdevelopment of firms. How can a state holding company, such as the OAE, develop such a strategy to intervene at the micro level for firms that lack the capacity for product design, marketing and the integration of planning and doing, and at the sectoral level where competitive market strategies may be mutually destructive or where intersectoral restructuring may be necessary? In chapter 3.2 we argued that much of the experience of the LEBs has suggested the need for planners and enterprises to work closely together in order to delineate the problems of particular firms and to provide packages of support for the need of firms, and to monitor the progress of firms. This does not necessarily entail the curtailing of firms' autonomy and the necessary flexibility entailed in operating in a competitive framework, or planners intervening in daily management decisions. This approach was often characterised as planning through markets. However the overall approach necessitates some clarity on the relationship between planners and new public enterprises.

How did PASOK's OAE strategy respond to this question? We can delineate two main conceptions of the appropriate relationship between the OAE and PFs. The first tended to emphasise the autonomy of PFs after their financial rehabilitation (Arsenis, 1985b; V Papandreou, 1986b). The emphasis here was on transforming PFs into autonomous, flexible and profitable enterprises, which would not rely on the previous, highly protectionist regime of the 1970s or on 'political' interference in day-to-day management decisions. V Papandreou stressed the importance of workers not seeing their inclusion in the OAE as a process of becoming civil servants with a guaranteed future, while Arsenis went as far as claiming that after their financial rehabilitation, the success of PFs would be dependent on good management.

This approach was indeed the dominant one within the OAE itself. In most firms 'budget committees' were set up to examine the financial difficulties of firms from the perspective of their budget and to identify ways of increasing turnover, improving efficiency, as well as proposing limited measures for restructuring. This in part reflected the fact that decisions on long-run viability were delayed and subject to political intervention from the government. This 'arms length' approach to the relationship between the OAE and the PFs, entailed the OAE in assessing the financial needs of PFs and undertaking a general supervisory, or monitoring role. This had some success in certain firms, in increasing their autonomy and flexibility. For instance, the largest textile firm was organisationally subdivided into ten smaller units to enhance product differentiation and a more coherent regional distribution of plants. Similar plans for other firms were worked out but had not been implemented by early 1988.

There is something to be said for this conception especially in the Greek context, where firms had traditionally operated within a highly protective framework, including political intervention in management decisions. Furthermore, given the nature of PASOK's post-1985 industrial strategy, the failure of the OAE to develop the infrastructure and even a conception for a more active planning role, this was not an irrational approach to the problems of PFs. However it clearly limited the role of the OAE to the rehabilitation of a few firms, and hardly constituted the incorporation of the OAE as a major mechanism in PASOK's industrial strategy.

The second conception of the appropriate relationship between the OAE and the PFs tended to support a more interventionist role for the OAE. Tsotsoros (1985), a Vice-President of the OAE, argued that the initial rehabilitation of the PFs could only be seen as the first, and in a sense easiest, part of the process. He claimed that the restructuring of the debt, while indispensable for the financial position of the banks, was unlikely by itself to provide a long-term viability for the PFs, many of which could easily return to a 'problematic' status if the OAE could not provide more long-term restructuring proposals. Tsotsoros argued that the OAE must go beyond an 'accounting' strategy to investigate financial packages and organisational measures needed to reorientate or diversify the activities of PFs and promote their international competitiveness. Merely changing the management of a particular firm, even if this led to an improvement in the quality of the personnel employed and asking this management to recommend a set of restructuring measures was unlikely to automatically affect the underlying problems of these firms.

Both the need for OAE to participate in overall industrial strategy and the long-run nature of the problems of PFs suggested to many that a somewhat closer relationship was necessary between the OAE and the PFs, providing OAE with an ongoing role in helping firms in their special investment, organisational and other needs. PFs would have to respond to both market forces and OAE incentives or planning guidelines. This conception is more in line with the type of strategic planning discussed in chapter three. It had supporters within the OAE itself. Thus while some PFs were dealt with the 'budget committee' approach, others within the OAE promoted a concept of 'internal hearings'. This entailed involving managers, workers and OAE officials, in order to examine the structural weaknesses which underlay financial problems. It was intended that such 'internal hearings' would lead to a final paper outlining both financial measures and longer-run measures for organisational restructuring.

In the practice of the OAE the tension between contending conceptions of its relationship with the PFs was never resolved. Indeed I was able to find no OAE policy document, even at the theoretical level, which stipulated in any detail how the OAE planners, PF managers and workers were to operate. In this context the nature of the OAE intervention tended to be <u>ad hoc</u>, the degree of progress depending on the contingent quality and abilities of particular managers within the PFs or the OAE staff, rather than the implementation of an agreed strategy. Nor is it surprising that of the two conceptions of the appropriate relationship between the OAE and the PFs, the more limited one, on the whole, was adopted. By 1987 the OAE at least began to close down non-viable firms, as well as to propose restructuring for the rest, which included redundancies - 6,800 lay-offs were envisaged, representing 23% of employment in PFs. In 1986 and 1987, many of the viable firms recorded improved turnover and profits, although this reflected in part the reduction in real wages of workers, as a result of the 1985 stabilisation programme (OECD, 1987, p35-6). Nevertheless OAE had saved a number of firms from certain liquidation and so the policy in this respect was partially successful. But the OAE was unable to progress to the implementation of its other aims of acting as a decentralised organ of industrial planning, searching for market gaps, setting up new investments or new public enterprises.

However, it is worth saying something more on the obstacles that existed for the broader conception of the role of the OAE and its relationship with the PFs. For instance, negotiations between the OAE and the new PF managements was hampered by the lack in most PFs of a basic institutional framework for their own corporate planning. Many of these firms lacked not just 'modern' management techniques but any mechanism for corporate planning, manpower planning or workers participation. Before a package could be discussed with the firms for long-run restructuring, the OAE had to encourage the creation of such mechanisms and institutions. What qualitative and rather sketchy evidence we have been able to gather suggests that the OAE was at least partially successful in this task. This suggests that the scale of the task confronting the OAE was considerably underestimated, given the time it would take firms, both managers and workers, to adjust to new institutional arrangements.

A second obstacle to the broader conception lies in the precise details of what is entailed in having a closer relationship between planners and new public enterprises. Our discussion in chapter three, especially with respect to LEBs, while indicative of a general approach, leaves a number of questions open, which we argued were subject to a lack of both theoretical analysis and practical experience. Of course there can be no one 'correct' theoretical position in this sphere. In our earlier discussion of the role of extending social ownership as part of a planning alternative, we noted the importance of clearly specifying the rationale for such intervention - whether the rationale was for purposes of promoting risky investments, introducing high technology or merely the gathering of information to aid sectoral planning and state incentives. Such considerations are also of relevance to the role of state holding companies and the extent of their intervention.

If a strategy of planning through markets is adopted then one resolution to the problem is that the market should be decisive for questions of product and taste, leaving planners to intervene in questions of investment, choice of technology and the link of a particular firm with other firms in a particular sector. In this perspective, the OAE could have been responsible for strategic matters to do with investment within the framework of the national plan. It is clearly important (see chapter 3) to decide which decisions should be taken at a central level. A good example of this was an OAE report on its two paper manufacturing firms, which suggested that great savings could be attained by promoting the specialisation of each firm in particular products, rather than having both producing the whole range of possible paper products. In other firms the extent of intervention from a state holding company can be far less, concentrating instead on supporting management decisions through financial packages, marketing advice and research and development services. Yet other firms can be run with full autonomy from state holding company intervention, although their role within the state holding company may be rationalised for purposes to do with the overall industrial strategy - providing information on sectoral conditions or assisting in technological diffusion.

Two conclusions follow from the above. That the successful intervention of any state holding company depends on such clarity on the goals and means for such intervention. All too often, left social democratic theory and practice has lumped such goals and means together, making it extremely difficult to articulate a policy which has clarified the scope of intervention and in which sectors or firms it is relevant. Secondly, that a successful relationship between planners and public enterprises depends on the overall institutional framework of industrial policy. In this respect the failure of the policies for sectoral industrial planning was a fatal flaw in the OAE's capacity to intervene in PFs, let alone extend its role in setting up new public enterprises. And this brings us to the question of administrative capacity and the flow of information, on which we have argued elsewhere, the whole coherence of a planning alternative, taken as a package, crucially depends. A more active role for the OAE would have necessitated a better articulation with other aspects of PASOK's planning alternative such as the national plan and sectoral industrial strategy.

Finally we should point to a further obstacle to the wider conception of OAE's role. For in many left social democratic accounts the role of workers participation plays an important role in providing the motor force to overcome many of the obstacles we have been examining - without such participation a left social democratic government is deprived of the vital knowledge and power that is necessary in any planning exercise. And as late as 1986, V Papandreou (1986b) was emphasising the importance of workers participation in the OAE strategy. She argued that workers should be involved in the whole range of management decisions, expanding their role gradually to questions of production and investment, and participation in the profits and shares of PFs. So what can be learnt from the OAE experience on the extent to which workers participation can assist the role of state holding companies?

Law 1386/83 limited the institutionalisation of labour participation to three areas: a union representative on the nine-member board of the OAE and the MNE Secretariat and one representative of the workers on the management boards of the PFs. As to the latter, Law 1386/83 did not specify the role or scope of workers' representatives in PFs, and the spirit of the law seems to be more the participation of labour as a factor of production than one of workers' participation with a more active role (see Mitropoulos, 1985, p84-8).[13]

Nor did the OAE itself attempt to institutionalise workers participation in the PFs, within the spirit of PASOK's frequently pronounced participatory philosophy. Again we were able to find no document outlining the role envisaged for workers in PFs. What improvements were made in the reorganisation of production, with respect to industrial relations, was therefore again reliant on the approach of particular OAE officials or PF managers. No systematic attempt was made to introduce measures for institutionalising workers participation or even, as a first step, discussing with workers the possible forms that such participation could take. Nor were any of the ideas of giving workers some participation in the shares and profits of PFs ever implemented.

We need not repeat the set of factors which prevent such workers participation in the Greek case (see chapter 6.3) such as the weakness of civil society, the organisational and ideological nature of Greek trade unions and the populist elements of PASOK's politics. Clearly these factors were also relevant here. However there existed a set of further problems for workers participation in the OAE strategy. The ASKE solution, we argued, promoted a relatively sophisticated response to the problems of extending social control to state enterprise by providing an alternative to both self-management and Morrisonian nationalisations. However this was facilitated by the fact that the ownership status of such firms was clearly recognised as that of state-owned enterprise.

The ownership status of PFs was decisively less clear, in part because of the political controversy over their future status, but also because of theoretical confusion of what the appropriate role of workers participation should be. In chapter 3.1 we noted the importance of the left social democratic model having a clear conception of the various forms of ownership possible and a strategy of what form was appropriate for particular enterprises. This was related to the question of at what level planning intervention, if any, was required, a theme we have been discussing over the relationship between the OAE and PFs. Such considerations also have a bearing on the appropriate level of workers participation. Here too, left social democratic models of planning have tended to lump together various arguments on the means and goals of strategy, which has not allowed a framework for rational promotion of workers participation. Only the existence of such clarity on the means and goals of strategy can allow progress towards providing a framework for workers participation, in turn allowing discussion, and practical experience, on the best methods of incorporating such participation. As our discussion on the ASKE clearly indicated even this stage is likely to be a complex one, entailing political conflict and an extended period of 'learning by doing'.

We have here examined three major obstacles to OAEs potential of developing into a more interventionist planning institution to assist industrial strategy: the existence of considerable administrative and organisational preconditions, and the lack of a clear conception on both the role of the OAE in relation to PFs and of workers participation in the OAE strategy.

7.4 Conclusions

In this chapter we have examined the limited success of PASOK's industrial strategy. A number of conclusions can be drawn on both the relevance of the alternative planning model for the Greek case, and on the lessons that can be drawn for the model itself from the PASOK experiment.

Broadly two set of arguments are relevant here. The first relates to PASOK's inability to harmonise the short and long-term aspects of its strategy. We have seen numerous examples of this phenomenon. The most significant is the delays involved in setting up the institutional basis for the OAE, the failure to reach decisions on the viability of PFs and the reluctance to accept the short-term costs of certain critical policy decisions, and the overall lack of a coherent development policy outlining central economic priorities.

This is linked to PASOK's failure to reach a 'global' strategy to mediate and harmonise the interests of various sections of PASOK's political constituency. Furthermore there was a lack of mechanisms, at either party or government level to reach consensus over policy, and a failure to institutionalise mechanisms for debate around its policies with the rest of society (see the branch studies). Such lack of co-ordination entailed that PASOK could not bring the various aspects of its strategy together as a package. We have seen the failure to link the OAE, SCs and branch studies into a coherent overall industrial strategy.

At one level, therefore, these failures reflect a contradiction between PASOK's interventionist strategy and its populist politics. In chapter two, when discussing the French indicative planning experience, we noted some possible causes for the decline of planning in the 1970s. One factor was that in the uncertain economic context, French governments were determined to allow themselves a considerable level of autonomy and flexibility in face of changing economic and political developments. It was argued that planning, if it enhances transparency in decision-making and a shift of decision making towards more long-run considerations tends to limit such autonomy. This was also a strong motivating force behind PASOK's lack of an industrial strategy - a more forceful attempt to introduce some of the mechanisms and institutions that PASOK had originally envisaged would have clearly limited PASOK's overall autonomy at the political level. This may well have contradicted its populist strategy of building up its social base.[14] The above arguments tend to focus on the 'political will' of PASOK, although they are nevertheless important in stipulating certain important political economy prerequisites for alternative economic strategies. However there exists a second set of arguments which attempt to go beyond the 'political will' explanation.

The first aspect of this is the considerable difficulties to be expected in setting up new institutions for planning. In this, and the previous chapter, we have seen PASOK's failure to establish the infrastructure and organisational framework for the operation of planning institutions on a long-term basis. PASOK seems to have severely underestimated the legal, administrative and organisational obstacles involved. Policy initiatives were continuously subject to such problems entailing delays that severely undermined the overall strategy and the goals of the national plan. Such considerations generate important doubts on the speed in which such a process can be initiated, and the size of the supply-side response, at least in the short run, on which many alternative economic strategies rely. This reflects not only the establishment of individual planning institutions but how they are coordinated together and how the problem of information flows between the new institutions and the planners is solved. On the resolution of such problems, we argued in chapter 3.5, the whole coherence of a planning alternative rests. It also relates to the question of comprehensiveness of the planning alternative, whether instead of attempting to set up a comprehensive set of institutions, it is not better to focus planning on priority areas, allowing the experiment to be expanded depending on its success and the level of support it generates.

However the question is not merely one of the speed and comprehensiveness of planning. For we argued in particular with respect to the OAE, that the coherence of new planning institutions also depends on resolving certain questions of their operation which are undertheorised in the literature and where practical experience has been limited. Thus with respect to the OAE we noted the lack of clarity on the appropriate relationship between planners and PFs, and on the role of workers participation. We argued that this phenomenon would have provided severe obstacles to the OAE developing a more interventionist stance, even had the 'political will' for such an experiment been forthcoming. Thus we can reverse the argument of 'political will' which emphasises PASOK's lack of a global strategy. For without a workable alternative on how workers, managers and the OAE could have worked together to promote a different conception of planning it was difficult to create such a global strategy to harmonise the interests of various social groups at the central level. Furthermore only the existence of such an alternative can increase the active support for such an approach among those groups that have something to gain from the introduction of planning. By looking at some of the obstacles to the OAE's planning function, we have delineated certain important prerequisites for the success of such a planning alternative.

Notes

1. Mouzelis (1986) provides a sophisticated dependency analysis and its relevance to Greece. An extensive and thorough examination of contending approaches towards Greek industrialisation in this period is provided by Petmetzidou-Tsoulouvi (1984).

2. Lyberaki's (1988) work on small-medium industrial firms in Greece argues that the lack of a flexible-specialisation strategy in Greek manufacture has been a significant factor in its decline. She also analyses the potential for such a strategy, basing her argument on some recent, although incomplete, flexible specialisation experiments among small and medium-sized producers.

3. For New Democracy's spasmodic attempt to deal with the 'ailing industries' phenomenon and the financial position of the banks, see Stergiou (1985). By 1979 the government had produced a number of measures, embodied in Law 876/79, which allowed for a partial capitalisation of 'ailing industries' debt, but nothing was ever implemented.

4. Law 1385 specified a number of measures to ensure that the information which could affect the competitiveness of firms was protected.

5. Marinos-Kouris, an academic at Athens Polytechnic, was the central figure in these working teams. Our account in this section is based on interviews with him and other members of the branch studies working teams.

6. From various interviews conducted with those responsible for the OAE process in this period, it became clear that there was considerable confusion surrounding the final criteria for firms entering the OAE. A number of firms entered merely because of their size and the fears of allowing their workforce to be laid-off while there seems to have been a considerable political bargaining between interested parties and 'informal pressure'. The process was not distinguished by any marked improvement in the

transparency of decision making, perhaps contributing to the inadequacy of the viability studies.

7. Panagopoulos (1985) considered the eventual decision of October 1985 as both realistic and just. He also distanced himself from the criticisms of the private sector, stating that it was just that the old owners should lose control of that part of the capital represented by the debt.

8. For examples of proposals favourable to the private sector approach see Kapanitsa (1986) and Lekatsa (1986).

9. This reflects the well-known problems faced by selective interventionist strategies, discussed in chapter 3.3 and 3.4.

10. In interviews with OAE personnel, it was a frequent complaint that this foothold of the old owners in the process (for instance to call general meetings and obtain information briefings at short notice) was often used to obstruct the rehabilitation process.

11. From interviews carried out it became clear that during this period (1985-87) there were broadly two positions. The first wanted OAE to keep 7 of the 21 viable PFs, with the rest being sold to the private sector. This implied a rather limited conception of the OAE's role in industrial policy, given that the seven 'strategic' firms were likely to be the least profitable. The second position wanted OAE to have a controlling stake in all 21 viable firms, although some shares could be distributed to workers, local authorities or small private investors. The latter position entailed a larger role for the OAE in industrial strategy.

12. PASOK claimed that Anomeritis resigned for 'personal reasons'. In fact it seems that while he did not object to the policy of privatising some PFs, he insisted that the funds generated by any sale should be ploughed back into OAE to enable it to break into new areas. The whole period is covered by extensive press speculation (see Augi, 22/4/87 and 23/4/87; Eleutherotipia, 30/4/87 and 18/9/87).

13. PASOK's promotion of workers participation within the OAE legislation was far less committed than it had been over the ASKE. Perhaps this reflects a fear that given the problematic nature of the firms involved, any failure of expanding participation would discredit participatory policies as a whole.

14. Similar considerations were seen to operate in PASOK's inability to promote 'stabilisation through development' (chapter five).

8 Conclusions: the political economy of alternative economic strategies

The significance of the PASOK experiment, at least during its first term (1981-85), lies in the fact that it constitutes an interesting test case of what we have termed the left social democratic approach to economic theory and policy. In its programmatic guidelines and ideology PASOK sought to initiate a 'third road' to socialism, rejecting both traditional social democracy and the revolutionary path. The promotion of institutional reforms such as democratic planning and socialisation, and new institutions or mechanisms for planning such as planning agreements or the OAE must be seen in this context.

From the analysis of chapters 5-7, it is clear that our conclusion is that PASOK was unsuccessful in this experiment. As we have seen the failure of the stabilisation through development strategy, both the inadequate adjustment from the demand side and the paucity of the supply-side response, led to a change in PASOK's direction after the 1985 election. Stabilisation of the economy became the number one priority with the announcement of the stabilisation package (devaluation, real wage cuts, targets for cutting public deficits and so on) in October 1985.[1] The structural or institutional reforms of the first period either remained at the level of pronouncements and were never implemented (supervisory councils, sectoral industrial strategy) or were promoted with little conviction and with a narrower scope (ASKE, OAE). Furthermore at the ideological level the emphasis was now on modernisation and efficiency and less on participation and transforming the balance of economic power within society. The new set of institutional or structural reforms aimed at the liberalisation of the banking system, 'flexibility' in the labour market and the promotion of venture capital and the stock market. It could be argued that these new priorities do not necessarily contradict an alternative economic strategy. And given the tradition of statist intervention in Greece, and the fact that, as we argued in chapters one and three, the left social democratic model does not imply a simple replacement of the market by the plan, there is something to be said for this view. However the virtual abandonment of the previous priorities, and in particular the lack of a coherent industrial strategy indicates a clear change in direction. Interestingly enough PASOK's post-1985 policy was hampered by problems similar to those confronted during its first term. Thus the inability to promote a global strategy and coordinate policy, as well as the politics of populism and clientelism led to severe contradictions. If PASOK managed to stabilise the economy to a certain extent (at least in the period 1985-87 - although public deficits were never brought under control), then some of these problems were actually accentuated. In particular the second term, and particularly after 1987, was bedeviled with allegations of widespread mismanagement and corruption among state ministers and officials. In this context in the last two years it was difficult to detect any coherent economic policy whether traditional or 'alternative'. Thus we do not examine PASOK's post-1985 policy but will discuss the nature of the failure during its first term.

Given the numerous obstacles that PASOK's strategy confronted and its inability to respond with a coherent and coordinated economic policy, the question arises whether PASOK failed because its goals and policy were in some sense unrealistic. Two strands of argument permeate the analysis of chapters 5-7. On the one hand, we have stressed certain political economy obstacles to PASOK's strategy. At this level it could be said that PASOK's failure reflects a failure of political will. But we have stressed that political will arguments can disguise as much as illuminate the important processes involved. At a more fundamental level PASOK's failure reflects its inability to delineate central priorities for its economic policy. Democratic planning, the institutional framework for extending democracy to the economic sphere, almost by definition implies an increase in the transparency of economic decision-making in which clear winners and losers will be indicated. In this light therefore PASOK's failure to delineate economic priorities also reflects its failure to harmonise the social and economic interests that it purported to represent and thereby promote a global strategy.

The second strand in the argument however points to the fact that the lack of a global strategy was not only a question of the absence of an appropriate 'politics' of planning. Here the emphasis was more on the technical and economic aspects of an appropriate form of state intervention, and for new planning institutions and mechanisms. While PASOK, at the programmatic level had a conception of instituting various planning and interventionist mechanisms and institutions, this did not extend to fully worked out operational plans of how these would function in practice. How comprehensive should planning be? In which areas, and on what rationale, should social ownership or planning agreements be promoted? What is the appropriate relationship between the state and the planners or between the planners and the individual firms and/or sectors of industry? What role should workers participation play? To such questions PASOK had at best a very hazy conception before coming to power with the result that its initiatives once in power tended to be spasmodic and uncoordinated.

It is only when these two strands are taken together that we can adequately address the realism of PASOK's goals. The ability to promote a realistic strategy depends on knowing the obstacles that are likely to be faced and how they can be overcome. This necessitates both a global strategy and a coherent understanding of the scope and potential for various proposed economic mechanisms and institutions. Only then can one assess the extent to which opposition to the policy can be overcome or the extent to which such opposition will have deleterious effects necessitating some form of compromise.

Thus we conclude here by delving deeper into PASOK's failure from these two strands of the argument. We discuss both the relevance of the left social democratic model for a country like

Greece, and conversely the extent to which PASOK's failure also points to inadequacies or gaps within the model as such. Does the PASOK experiment have wider implications for future alternative economic strategies?

8.1 Planning, state and society

In chapters six and seven we discussed a number of obstacles and contradictions in PASOK's strategy of setting up a system of planning to promote the supply-side response of the economy. We examined the organisational, administrative and political obstacles for both the framework of democratic planning and individual institutions or mechanisms such as planning agreements, the OAE and supervisory councils. Here we shall not restate these conclusions (see chapter 6.4 and 7.4). We further argued that the organisational, administrative and political obstacles to PASOK's strategy were all to a degree socially determined. Thus in this section we develop this argument by examining the relationship between PASOK's planning proposals and the nature of the Greek social formation and the state. For while we believe the nature of the Greek state and social formation provided a particularly infertile ground for the implementation of a planning alternative, PASOK's confrontation with certain problems does raise questions of wider interest for any left social democratic strategy.

8.1.1 Planning and society

There are a number of aspects of the Greek state and social formation which have a bearing on the fate of PASOK's planning alternative. Some have already been alluded to in chapters 5-7. In chapter 7.1 we argued that the Greek state has been crucial in post-war economic development. While this did not entail any coherent development strategy, it did imply the state's involvement in all aspects of economic activity. This had important consequences for the nature of the private sector. Thus even as late as 1986, the Bank of Greece (1986, p23) was critical of the policy of successive governments which had led to a private sector concentrating its activities on expanding the degree of protection or level of subsidy rather than relying on its own dynamism to improve its competitiveness or restructure its production.[2]

It is not, however, only the private sector that seeks such a relationship with the state. Rather it extends to a wide range of social groups or classes, with the result that the financial system and state subsidies have been used more as instruments of controlling income distribution than as development instruments. This reverses the usual logic of liberal-capitalist economies, where activity in civil society, the ownership of means of production or private property, determine political power. In Greece, it is access to political power which is often the determinant thereby severely reducing the ideology of profit and competitiveness (see Petmetzidou-Tsoulouvi, 1987, pp199-200).

The importance of access to state power is the framework in which to understand such recurring themes in this book as clientelism, ad hoc government interventions and PASOK's inability to set a different course. It is clearly beyond our scope here to examine the historical development of the Greek state and social formation (see Tsoukalas, 1986; Petmetzidou-Tsoulouvi, 1984, 1987). Rather we seek to explore how this affected PASOK's planning proposals.

Xtouris (1987) has attempted a critical analysis of the relationship between some of the above themes and the role of various state economic and planning services at the local (Nomos) level. He notes first that at the Nomos level these services have a myriad of legislative and administrative instruments at their disposal, with the result that most forms of economic activity can be encompassed in the existing legal framework. However the quality of these services suffers from a number of inadequacies: the fact that the legislative framework has been built up over the years with virtually no attempt to make it consistent and internally coherent; the lack of training of local state officials; the extent to which responsibility is spread over a large number of planning organs and institutions with the result that few know their precise role or responsibility and how this fits into the overall framework (Xtouris, 1987, p46).

The major result of this is that any form of planning becomes extremely precarious. For any economic activity or investment entails a certain degree of forecasting and security. In the Greek case it is the lack of such security which leads to the importance of access to state power. For clientelism, personal relationships and even corruption can ease access to state subsidies, licenses or other forms of protection.[3] As Xtouris (1987, p40) concludes this phenomenon considerably weakens the ability of the state to promote development and a clear set of rules for the operation of economic activities. Given the extent of state aid that is mediated through personal or clientelistic relationships, it is virtually impossible to work out the real costs of these incentives and thus promote a rational policy for the future (Lambrianidis, 1987, p30). Needless to say such intransparency systematically operates for the benefit of certain privileged economic and social groups.[4]

The lack of an infrastructure for planning at both the local and national level must be set in this context. At the local level it explains the poor quality of state personnel, the difficulties entailed in demarcating clear roles and spheres of responsibility and the absence of adequate research and other facilities (Lambrianidis, 1987, p31). At the national level there are similar problems as we have seen. Thus in chapter 6.1 we saw that a proper framework for democratic planning was hampered by such phenomena as the lack of institutions for coordination of policy at the government or society level, the continual <u>ad hoc</u> ministerial interventions in the formulation of the national plan and the work of KEPE, and so on.

As both Xtouris (1987) and Lambrianidis (1987) have argued, PASOK continually wavered between a recognition of the costs of unplanned development in terms of the rise of the black economy, public deficits and a structural balance of payments crisis, and a fear of taking any decisive action to address this which would have political costs. For from the above it is clear that significant social groups, and not just the private sector, benefited from this lack of planning. Furthermore many of these groups, and especially the middle classes, formed a natural constituency for PASOK.[5] If such groups would support certain modernisation proposals promoted by PASOK this did not extend to a renunciation of the special privileges and access to state power they had previously acquired (Xtouris, 1987). As Lambrianidis (1987, pp32-3) concludes planning failed in Greece because it was not built on an alliance of all those who suffered from its previous absence.

Thus the failure to promote a planning infrastructure does not merely reflect administrative and other more technical factors. This is a first conclusion of more general relevance. As

Dearlove and White (1987, p2) have argued it is particularly important to analyse:

> the political nature of <u>economic reform coalitions</u> and the relation between economic reform 'from above' (state sponsored) and 'from below' (socially demanded). One may well doubt the capacity of state elites to transform themselves and undermine the basis of their own power and privilege (unless in unavoidable dire straits). The political basis of economic restructuring will be stronger if there is the possibility of alliances between reform-minded elements of the established regime and those social interests from below that are dissatisfied with the status quo and set to benefit from reform.[6]

This clarifies the manner in which planning necessitates appropriate 'politics' which is socially determined. However, PASOK was unable to promote such politics. As we saw in chapter 4.1, PASOK's political strategy was based on the nebulous formulation of the alliance of 'non-privileged Greeks'. Such a formulation was singularly unable to tackle the contradictions within this alliance when PASOK came to government. The failure to set up an infrastructure for planning, to co-ordinate policy initiatives and delineate economic priorities must be seen this light.

In chapter 4.1 we noted that PASOK's strategy rested on a series of structural/institutional interventions and that PASOK, in opposition at least, was aware of the importance of the rate or speed of their introduction - a very slow rate would risk their incorporation into the existing socio-economic system while a too high rate would entail dislocation or disorganisation of economic activities. However in government PASOK seems to have interpreted this sensible conception in a peculiarly populist manner which gave little operational guidance to what could and should be achieved. The conception was reinterpreted to mean that PASOK should not go further than the 'people' were willing to accept. The question arose of how the will of the 'people' could be ascertained. The resolution to this dilemma was that PASOK, because it lacked a framework for resolving differences and coordinating policy either at the level of party or government, responded to organised expression of opinion which was hostile to individual measures. Moreover, this process usually occurred 'behind closed doors'. This led to a great deal of delays and uncertainty on the eventual nature and scope of various institutional interventions (see for instance our discussion of the policies for supervisory councils and the OAE in chapter seven). This was also the case for other aspects of PASOK's strategy. For instance in chapter five we saw that PASOK's ability to promote macroeconomic control was severely debilitated by its inability to restructure the taxation system. The failure to present a new initiative on tax evasion and the taxation of farmers and the self-employed was the result of pressures on PASOK from elements within PASOK's political machine and its natural electoral constituency.

For in chapter one we argued that a left social democratic strategy could not rest merely on aggregating all the various short-term demands from its natural constituency but must seek to harmonise the short and long-term interests of that constituency to promote a global strategy. PASOK as a political party, and the nature of its electoral constituency, was particularly unsuited for this task. Thus its strategy was unable to exploit the 'mobilising dynamic' of structural interventions which could have united all those who had an interest in the transformation of the social-economic system. Rather PASOK in its first term concentrated on the 'easy' aspects of its policy - wage increases, higher pensions and social policy (Valden, 1986).[7] On the other hand the structural or institutional measures, under the pressures of opposition from both within PASOK and forces hostile to it, were subject to, at best, spasmodic and uncoordinated promotion thereby considerably weakening their mobilising dynamic. Thus while not belittling some of PASOK's achievements in the social sphere, especially the creation of a National Health Service, these were initiatives that any reforming administration could have pursued. But PASOK's original conception (chapter 4.1), mirroring that of left social democracy, was that such reforms were neither cumulative nor irreversible unless accompanied by structural reforms.

However as we have also argued all this was not only the result of PASOK's shortcomings, for any party must to a certain extent reflect the society from which it arises. We have stressed in this connection the importance of the weakness of civil society in the Greek social formation. For instance the conception of a decentralised and local planning is debilitated by such weaknesses. The lack of an autonomous trade union movement, in the case of the OAE and ASKE, considerably hampers its ability to promote its own strategy and thus participate in a transformed institutional setting for participation in national planning and individual enterprises. While such factors were particularly determinant in the Greek case, they would not be absent in any other left social democratic experiment.

8.1.2 Planning and the state

The emphasis on the political economy prerequisites for a planning alternative should not obscure the second strand of the argument that has permeated our analysis. That is that without a clear conception on the details, economic rationale and technocratic support of the planning alternative it is impossible to promote the political economy prerequisites. Here there are even more implications for some of the ambiguities and inadequacies of the left social democratic model analysed in chapter three. Let us begin with the role of the state in any planning alternative.

The problems of state intervention in Greece before 1981 and PASOK's failure to restructure the state to make it more suitable for a new form of intervention point to important conclusions for the coherence of the left social democratic model. To many they would point to the inherent limitations of the state with respect to economic activities and the superiority of market methods. The backlash, within the economics profession but also in society at large, against various forms of state intervention (central planning, macroeconomic demand management and indicative planning, and developmental states in the Third World) are indicative of this. As Dearlove and White (1987, p1) point out the problem is thought to rest not only in the nature of state-sponsored policies but with the state itself - the incapacity of state officials to manage the economy efficiently, the distortions in development induced by state intervention and the fact that the domination of the state is often tied to the emergence of new classes/strata whose power is rooted in political or administrative office. In the economics literature such phenomenon are analysed in what has come to be known as public choice theory and the 'new' economics of politics approach.

Our account of PASOK in government provides a certain

credence to this widespread revaluation of state failure as opposed to market failure which is the basis of much left economic policy. Thus in chapter 6.2.1 we saw PASOK's inability to promote the large industrial projects which it had supported. When these failed to materialise PASOK was left without a conception of the type of development it was intending to promote. It was uncertain whether its central task was to foster such large state industrial projects, to support small or medium-sized firms, to emphasise the large infrastructural projects such as the Athens metro or the Achelos dam, or to promote the service sectors (tourism, banking etc) of the Greek economy. The later emphasis on the large infrastructural projects it could be argued is another indication of state failure, and the megalomania and desire to impress of state officials. In the end, the state, under PASOK, dallied with all these forms of state promotion of development and was not particularly successful in any.

And yet PASOK was initially aware of the problems of state failure. As we have seen its support for democratic planning and the three major institutional reforms, codified in three corresponding laws (OAE, Supervisory Councils and the socialisation of the DEKO) reflected its belief in the inadequacy of the existing public and private sectors. However it was only with socialisation and the ASKE that PASOK was able to sponsor some genuinely new thinking on how to promote some form of social control in the public sector which went beyond Morrisonian nationalisation and self-management models (chapter 6.3). The force of this change was lost because it constituted such an isolated example and was not coordinated with other initiatives. In this context it could foster little political support even from those, such as the workers in the ASKE, who presumably had most to gain from the new institutional setting.

Furthermore in this case, as in all other areas, PASOK initiatives were crucially weakened by the role of clientelistic practices in the appointment of officials, the use of the new institutions for PASOK's social policy and its desire to build its electoral and social base, and <u>ad hoc</u> ministerial intervention at all levels. Thus a central contradiction in PASOK's approach was its inability to reform the public administration as it had promised. Indeed its support for new institutions for planning had as a consequence the postponement of a reform of the existing public administration. The result of this was that instead of these new institutions developing new initiatives and administrative practices, they merely reproduced the practices and modes of operation of the existing public administration.

PASOK's initial response to the public administration was to set up parallel organisations of political advisors to assist ministers and other state officials. While this might have been necessary in the short run, it quickly deteriorated into PASOK cadres intervening in all aspects of policy formation in a clientelistic, intransparent and uncoordinated manner (see Koulouglou, 1986, pp132-6). Rather than a democratisation of the state, what happened was a 'statisation of the party' with nearly all members of PASOK's central committee being coopted into either the government administration or parallel advisory organs (Koulouglou, 1986, p136). PASOK's failure to transform the state machine is also reflected in its centralising dynamic which quickly overtook its participatory rhetoric, in the poverty of its decentralisation measures (see chapter 6.1) and its abandonment of its support for self-management (see Spourdalakis, 1986, pp260-2).[8]

All these examples of state failure confront any left social democratic strategy with important questions. Is it the case, as the new orthodoxy suggests, that only by widening the scope of the market sphere can such failures be rectified? Or is there an alternative approach to confront both state and market failure? For it is clear for the reasons discussed in chapters 1-3, that the marketisation of economic activities creates its own set of problems which are incompatible with any left social democratic strategy. We argued there that markets and market actors are themselves politicised and that the centrality of politics cuts across the state-market divide (see also Dearlove and White, 1987, pp1-2). Furthermore markets themselves create their own inequalities, concentrating power and influence in private hands. Economic policy then has to confront the unequal power and advantage in the existing market structure, severely limiting such goals as equality and the extension of democracy to the economic sphere.

Brett (1987, p35) proposes a way to go beyond a simplistic division between state and market: 'Central to the concerns of both market and statist theorising is the problem of <u>monopoly</u> - that is, the concentration of control over resources (including most importantly, control over the ability of individuals to exercise preferences in both production and consumption) in the hands of minorities. Both regard such control as problematic but attribute it to different sources and attempt to deal with it through different processes'.

To those who are still aware of market failures or wish to promote democratic procedures in economic activities through state, or other forms of social or collective processes, this conception has important consequences.[9] As Brett (1987, p36) argues there is a need for:

> fundamental re-evaluation of the discipline of public administration, long associated with the mechanistic and formalistic developments of classical Weberian principles of formally rational bureaucratic organisation. Instead we need a radical analysis of the way in which public authority can be organised in order to eliminate the barriers behind which officialdom has always been able to hide, to ensure that they are rewarded for performance and not simply for occupying positions and doing what they are told, and to guarantee that they will not use the monopoly power derived from the state to exclude creative alternatives which might threaten their readiness to go on doing the things that they find most easy and comfortable.

Clearly this is not a new question. Indeed the goal of transforming the state and public administration is as old as the socialist movement itself. Given its importance one can only be startled by the lack of concrete proposals in this area. Here we can reverse the argument of PASOK's failure which is based on state mismanagement and clientelistic politics. For PASOK had a very limited pool of ideas and proposals from which to draw upon to restructure the public administration, to promote genuine democratisation, accountability and devolution of power. Its failure therefore reflects not just national contingent factors but the inadequacy of the left social democratic model to confront such issues. This points to a major theoretical and policy agenda for future research and 'an eclectic fusion of ideas and practices from political science, organisation theory, business and management studies and public sector economics. The central focus would be on the question of <u>restructuring the state</u>, devising new institutional forms and methods of intervention which may serve to reduce its bureaucratic power and well-established developmental deficiencies' (Dearlove and

White, 1987, p3). The goal should be to restructure both state intervention and promote <u>genuine</u> forms of economic competition. However the level of theory and practical policies of what this would mean in practice is at such a rudimentary level that it constitutes a major contradiction in the coherence of alternative planning proposals.

Before going on to other issues it may be useful to give a concrete example of what is now necessary. Throughout this book we have suggested a certain potential for the flexible-specialisation paradigm of development and its articulation with strategic planning. Lyberaki (1988) has been one of the first works to discuss the potential for such a conception in the Greek context. One of her conclusions is particularly relevant for our purposes here, for she argues that strategic planning entails a flexible state in order to coordinate and respond to the needs of systemic flexibility: 'So, systemic flexibility embraces the state structure and functions as well, and thus the latter's relations to industry must be (radically) redefined' (Lyberaki, 1988, chapter 10.5). The notion of a flexible state is here juxtaposed to that of an inflexible state (extended hierarchical structure, strong centralisation, clear division of labour between innovators and operators and between departments and functions, and a tendency towards vertical integration so that the whole bureaucratic chain remains under a single command). We can make two conclusions which follow from Lyberaki's discussion. Firstly, in Greece there was not even a developed inflexible state, let alone a flexible one. But secondly, and of more general relevance, the details of how such a flexible state would operate, its scope and mode of operation and the level of its accountability to both consumers and producers, is extremely unclear. Those who support the strategic planning paradigm have not, as yet, gone on to consider the problems involved and thus promote policies of arriving at a flexible state. Once more, then the remaining agenda of appropriate forms for integrating state, planning and society is a large one.

8.1.3 Institutions and mechanisms of planning

PASOK's failure to promote a planning alternative also points to a second area of difficulty where the problems confronted reflected not only the contingent circumstances of the Greek case but gaps or inadequacies in the original left social democratic model. We have suggested that there were a number of areas in which the task in hand was hindered by a lack of clarity on both the goals and operational details of various planning mechanisms or institutions.

Thus in chapter 7.3 we argued that even if a political will had existed for expanding the role of the OAE into a major arm of PASOK's industrial strategy this would not have been an easy goal to achieve in practice. We looked at three distinct areas in this respect: the relationship between the state and the planners, the relationship between the planners and individual problematic firms (PFs), and the role of worker's participation. In all three cases the success of the OAE would have hinged on resolving certain questions which are undertheorised in the existing literature and where practical experience has been limited. PASOK was unable to give operational answers to many of the questions which arose from our discussion in chapter three: at what level (national, local) is intervention most desirable or feasible? How is a particular planning institution articulated with other economic policies or instruments? What is the appropriate balance between planning and the market? To what extent can active workers participation promote the efficiency or organisational restructuring of an industrial sector or individual enterprises?

To take the planning-market nexus first. It was clear from the beginning that PASOK's OAE strategy entailed the creation of enterprises whose success would depend on their ability to compete in the market. However the OAE was responsible for promoting the reorganisation of PFs and providing a number of other services to help their competitiveness (market research, R & D and so on). Furthermore it was sometimes suggested that the PFs could help in the overall restructuring of the industrial sector in which they operated. Therefore PFs would have to respond to both the planners' dictate and market forces. If this entailed a conception of strategic planning, a form of planning through the market as discussed in chapter 3.1, then PASOK had only a very hazy idea of what this could mean in practice.

PASOK's early theoretical schema of 'centre-periphery' and 'anti-monopoly' were progressively abandoned but their legacy was that PASOK lacked the theoretical apparatus with which to tackle such questions. The 'anti-monopoly' schema had two important consequences. It suggested that the obstacles to development rested in the dominance of a relatively small number of firms. Secondly, it provided the promise that a more desirable pattern of development could be promoted by the relatively simple expedient of taking over such firms and thereby providing control over the process of capital accumulation (Doxiades, 1984, p47).

As we argued in chapter 3.2.2 such an approach is fatally flawed. It is in part based on the idea that the preconditions for planning have been provided by the increasing tendency to plan in capitalist economies and capitalist firms. However, as we saw, this overlooks the fact that such planning operates within the framework of a market existing 'on the outside' and the need for some form of market even in a socialist economy. Even if firms are taken over by a socialist government, the organisation and direction of such firms does not entail a simple administrative or technical exercise. The practical arrangements over the appropriate market-plan nexus have still to be considered.

PASOK is not the only recent socialist party to have stumbled in this area. Hall (1986, p204) has pointed to the vagueness over the purpose of nationalisations under the French socialists. One of the goals was clearly to provide finance to state industries to ensure a relatively high level of investment in the economy. Beyond this there was severe disagreement both within the Socialist party, and between it and the Communists, on the role of this expanded nationalised sector. As the government faced increasing problems, the difference between formal ownership and a capacity to direct enterprises to achieve certain goals once more reasserted itself: 'The French state still had to clarify its goals and learn how to use such enterprises to implement them. In 1982-83 these companies had the worst of both worlds: they lacked adequate direction from the state to guide their long-term strategy, yet they were subject to sporadic government intervention into their daily operations' (Hall, 1986, pp205).

Although PASOK's strategy did not encompass the nationalisation of profitable private sector firms, its difficulty with its OAE strategy reflects a similar phenomenon. In the case of the socialisation of the DEKO, PASOK was also unable to promote the integration of nationalised enterprises into national planning. Although it did provide some kind of framework for social control in these enterprises, it eventually concentrated on measures for financial and organisational restructuring of the DEKO on a firm by firm basis. The inability of various socialist governments, in different national contexts, to promote the integration of nationalised industries into a planning framework clearly suggests the lack of an adequate conception of the goals and operational details of an alternative strategy in this area.

In chapter 3.2.1 we noted that there are a number of possible rationales for the extension of social ownership or the introduction of planning. However we also pointed to the paramount importance of specifying the areas in which particular arguments are relevant in order to be able to concentrate effort in such areas, to examine the scope and at what level intervention is most appropriate and to clarify whether, given the problems of control and capacity, social control cannot be better operated by more indirect means. We saw that the tendency has been to jumble together these issues thereby further inhibiting clarity. PASOK's experience with planning agreements and Law 1262 reproduces this unfortunate tendency. The goal of planning agreements was to incorporate the private sector into the national plan and rationalise the system of incentives given to the private sector. In actual fact the private sector faced the worst of both worlds: on the one hand an adequate system for planning agreements was not initiated which could have promoted some form of integration of the private sector into national planning; on the other hand the delays and uncertainties with this mechanism, as with others, failed to clarify the 'rules of the game' for the private sector. Once more it is clear that PASOK, lacking a clear conception of the type of development it wished to further, was unable to concentrate its promotion of planning agreements on priority economic sectors or activities. This was also a fatal flaw in PASOK's sectoral industrial strategy and the supervisory councils.

This brings us to the question of the appropriate comprehensiveness of planning. We argued in chapter 3.5 that for a number of reasons planning should not seek to be too comprehensive. PASOK supported a large number of new planning mechanisms and institutions but was unable to fully implement many of these and, more importantly, could not link the various aspects of its strategy into a coherent whole. Our arguments in chapter 3.5 suggest that PASOK would have been more successful if it had focused on a few carefully selected priority economic sectors or economic activities. If it had been able to delineate certain industrial priorities, it may have been able to integrate the work of the OAE, Supervisory Councils and its sectoral industrial strategy. As Doxiades (1984, p147) points out the lack of new investments in Greece, and especially in new products, reflects the poverty of the overall institutional framework including the necessary trained manpower, technical expertise, technology networks, guaranteed markets or commercial networks, energy infrastructure and so on. While PASOK could not have been expected to set up such a framework for all sectors of the Greek economy, given the administrative, organisational and legal problems which we have discussed, it could have done a lot more if it had concentrated in certain areas. Such an approach could have provided a certain amount of practical experience on which to build on as well as led to a growth in the support for such planning if it proved successful. It may also have provided potential answers to certain unresolved questions such as the manner in which various policy measures can be coordinated or ways in which flows of information between planners, the state and individual firms or industrial sectors can be facilitated.

8.2 Macroeconomic policy and structural change

In the final sections of chapter three we showed that the coherence of any left social democratic alternative would rest on the ability to harness macroeconomic or short-term economic policy to planning. PASOK's strategy here is of general interest because, apart from a brief flirtation with Keynesian reflation, it did attempt to promote such a coordination of supply-side and demand-side policies through its strategy of 'stabilisation through development' and 'gradual adjustment' (chapter five).

However we saw that in practice PASOK was unable to implement this complex package and that the degree of coordination and control entailed by the idea of gradual adjustment was lacking. We do not need to elaborate on the major conclusions of chapter five: PASOK's failure to institutionalise an incomes policy or provide a basis for a 'political exchange' conception; the corresponding problem of harmonising short and long-term interests; the effects of the operation of an electoral business cycle and PASOK's desire to build up its social base. At the theoretical level we argued that PASOK severely underestimated the degree of adjustment that was necessary from the demand side. There was considerable ambiguity over whether the strategy entailed reducing the net PSBR as such or merely restructuring public expenditure towards investment. Furthermore it was clear that PASOK did not fully appreciate the extent to which its incomes policy had significant effects, at least in the short-run, on the competitiveness of the Greek economy.

We further argued that such ambiguities also reflected an underestimation of the effect of public deficits on general economic stability, as well as posing limits on the resources available for the various supply-side policies. The result was not only the severe economic imbalances of 1985 but also the lack of resources for planning, and new investments. In short, PASOK could not provide a supply-side 'friendly' macroeconomic policy.

A similar conclusion has been drawn by Hall in his discussion of the economic policy of the French socialists in the early 1980s. Hall (1986, pp196-7) argues that a supply-side 'friendly' macroeconomic policy in France was hindered by the long-term changes in France's international economic position, and the growing interdependence of world production and consequently trade, energy and monetary policies. This severely limited the autonomy of French macroeconomic policy. The need for stabilisation of the Greek economy in both the pre-1985 era and beyond does reflect this phenomena and the need to synchronise the Greek economy with other European economies. Given the importance of liberalisation of trade and the internationalisation of capital flows, there is a need for a certain amount of policy convergence. To the extent that economic indicators do diverge, this necessitates corrective measures.

We can draw two major conclusions of more general relevance from this analysis. The first issue concerns the coming of the internal European market in 1992 which provides a new challenge for any future alternative economic strategy. There are two promising avenues of research in this context. The first is to examine the implications of the internal market for national economic policy making. The bulk of the existing theoretical work has focused on the expected benefits of the abolition of frontiers, the removal of capital controls and other barriers to trade (see, for example, Cecchini, 1988). Much of the analysis seems to assume that the European market will generate macroeconomic balance (full-employment, stable prices and regional balance) and promote the necessary economic restructuring. However, the problems of market failures and control over the restructuring process are likely to be as crucial an element at the European level as they have been at the national. There is a need to investigate the scope of national governments to intervene in their economies to promote growth, development and employment. To what extent is the loss of traditional instruments of economic policy (for example, public procurements policy and exchange controls) critical to undermining the ability of national governments to achieve the above objectives? If such a loss of autonomy is an integral part of the new economic context, are there any new set of economic policy instruments which can promote these objectives?

The second avenue of research follows closely from this. If national autonomy is severely curtailed what policy instruments or new institutions can promote the achievement of the above objectives at the European level? There have been a number of proposals from social democratic/socialist theorists on the promotion of the social element of the internal market, European-level reflation and economic restructuring (see, for example, Holland, 1983). However so far these have been more at the level of desirable objectives rather than fully articulated research or proposals on the new European institutions and/or policy instruments.[10]

However our second major conclusion suggests that we need not be so fatalistic on the extent to which national governments can carry out their own more particular macroeconomic policies and coordinate these with supply-side policies. For the failure of PASOK in this area reflects a different set of factors from its failure in planning. For PASOK did have both a relatively sophisticated conception and the necessary policy instruments for gearing macroeconomic policy to its supply-side strategy. Our analysis suggests the importance and interrelationships entailed in linking macroeconomic policy with industrial and social policies.

This issue can be clarified by briefly comparing our analysis of the Greek case with the experience of France and Britain. Tomlinson (1988, p9) considers the failure of reflation in Britain between 1974 and 1976. He notes the existence of some general obstacles to this reflation including Britain's weak competitive position in international markets and the absence of any social control over financial and capital flows. However he concludes that: 'Even given these constraints it is possible to argue that a policy of reflation which was coupled with an effective incomes policy, focused on areas other than the public sector wage bill and which was less ambitious <u>initially</u> might have prevented the 'boom and bust' sequence that occurred' (Tomlinson, 1988, p9). Critical to his argument is that such an alternative course was ruled out more by certain political conditions[11] and the 'nexus of union/labour relations, the ideologies and practices of the left, than with international economic integration <u>per se</u>'.

Similar conditions operated in France after 1981. The socialist government was committed to certain social policies, reduction of unemployment and income redistribution (Hall, 1986, p197). As in Greece, these policies led to a sharp rise in the deficits of social welfare funds and the national budget, and the rise in wages. The combined effect of this was that reflation was not 'supply-side' friendly (Tomlinson, 1988, p11). Once more the international environment was certainly a force limiting the autonomy of the government but: 'Much of the problem in both countries [Britain, France] lay as much in the internal policies of the ruling group and its allies, as in the external economic environment' (Tomlinson, 1988, p12). Hall (1986, p221) points to certain important forces operating in France in this context: the uneasy relationship between trade unions and the government; the failure of the government to discuss its macroeconomic policy with the unions; and the effect of interunion rivalry which restricted their ability to promote their own strategy. We have noted the effect of similar forces in the Greek case throughout this book.

Tomlinson (1988, p13) usefully compares the British and French cases with that of Sweden. He notes the extent to which in Sweden macroeconomic policy was both geared to supply-side policies and could, most importantly, generate a consensus within society. Thus wage restraint was more acceptable because it was associated with labour market programmes and a certain socialisation of investment through wage-earner funds - unions were more confident that they could expect a share in the benefits of their restraint. Secondly macroeconomic policy was operated in a context of certain amount of socialisation with respect to capital/financial flows. The existence of strong pension funds mitigated the possibility of reflation leading to a loss of confidence or capital flight. As Tomlinson (1988, p13) concludes: 'What the Swedish example shows is the interdependence of industrial, macroeconomic and 'social' policies'. The existence of such an interdependence has been given considerable support throughout this book. It suggests that while such problems an the internationalisation of trade and financial markets do set new constraints on alternative economic strategies, there is no need to accept these fatalistically. Rather the interdependence of industrial, macroeconomic and social policies underlines the importance of harmonising short and long-term aspects of any strategy and maximising the level of consensus and support for any alternative economic strategy.

8.3 Methodological issues

Our theoretical stance in this book has been to attempt to integrate the literature of economics, planning and development with what we have termed the political economy perspective. We can conclude here by examining advantages and disadvantages of both perspectives, while at the same time underlying the existence of the need for further research.

To take political economy first. The major advantage is that political economy sheds light on some important aspects of economic activities and policy-making which are hidden by the more traditional literature. It provides a framework for understanding the reasons why economists disagree amongst themselves, and can clarify some of the central obstacles to economic policies. The failure of planning in Greece reflects the fact that PASOK while accepting the economic costs of an unplanned development, was unwilling to challenge those classes or social groups which gained from this lack of planning. We have indicated that institutional transformation is most likely to be successful when reform 'from above' is allied to social groups most likely to benefit from a new institutional framework. With respect to a successful macroeconomic policy we have demonstrated, both at the theoretical level and in our analysis of PASOK's policy, the interdependence of macroeconomic, industrial and social policies. We have therefore pointed to certain political economy prerequisites at the level of the state and civil society, for a successful alternative economic strategy.

The disadvantages of the political economy perspective relates to the tendency to follow a course of analysis at higher and higher levels of abstraction. From an examination of a particular institution of planning, we go on to examine the social forces that are articulated with such an institution and then to various theories of the state. What is lost in this process is a detailed examination of the coherence and capacity of individual mechanisms or institutions to promote certain goals and how any one institution or mechanism fits into an overall economic package. Thus we have argued that PASOK's failure did not only reflect the lack of political economy prerequisites. Rather we argued that the gaps and inadequacies in the left social democratic model, severely restricted PASOK's ability to promote a global strategy.

The disadvantages of the more traditional literature on economics, planning and development follow from the advantages of the political economy perspective. However their advantage is that they facilitate a better framework for understanding concrete policy proposals. In this respect we have pointed to a number of areas where the development of this literature could contribute to a resolution to certain obstacles to any future alternative economic strategy. In particular we have argued that the ways in which the public administration and state institutions can be transformed to facilitate planning (accountability and the reduction of monopoly power rooted in the administration) is of critical importance to the overall coherence of alternative planning models. We have also shown the existence of certain gaps or inadequacies on the theory and implementation of individual planning mechanisms and institutions.

In the end, of course, Greece is a relatively small country whose experience cannot be expected to have too many lessons of wider relevance. We have tried to show, however, the extent to which the failure of the Greek alternative economic strategy was not due only to contingent national circumstances, and have drawn certain limited parallels with experiments in other countries. The possibilities entailed in combining a political economy analysis with the development of the existing literature on planning and development does therefore point to the existence of a large agenda for future research.

Notes

1. For an account of the rationale and details of the stabilisation package, see Ministry of National Economy (1985).

2. For an interesting article on the economics of this phenomenon, which employs analytic concepts first used in the context of East European economies (such as the notion of 'soft budget' constraints) see Katseli (1990).

3. The benefits that this privileged access can hope to achieve are numerous: urban planning permission, exemption from pollution controls, tax exemptions, direct income transfers, special licences etc (Lambrianidis, 1987, p30).

4. It will be noted that there is a similarity in this analysis with that often made with central planning models. As many have pointed out the level of corruption, black economy and personal relationships in the Soviet Union reflect the gaps in the centralised planning system. Political relationships can mitigate these inadequacies (see Hodgson, 1984, pp102-3).

5. Vergopoulos (1986a) argues that in the 1960s and 1970s certain middle-class groups gained considerably at the economic level from the nature of post-war economic development. However they felt that they lacked a corresponding power at the political level and felt excluded from the dominant political processes. Vergopoulos argues that PASOK attracted many of this group to its lines under the slogan of Allaghi (change). Clearly a sociological account of the class nature of PASOK's supporters or members is beyond the scope of this book.

6. A similar approach can be detected in Gilhespy et al (1986, pp12-13) and their proposals for an alternative economic policy. They argue that in Britain influential groups in the economy were unable or unwilling to use their power constructively and that no institution or group

had the power to promote effective long-term planning. Thus their alternative rests on a strategy for political and social change based on a collective approach to economic problems 'leading to the enfranchisement of all those whose needs the economy should be meeting'. This reflects a widespread new thinking on the left for the role of 'empowerment' in any alternative strategy.

7. Kesselman (1986, p235) provides a similar analysis for the fate of the socialist party in France after 1981:

> The first year's reforms changed many aspects of France's political economy. But rather than generating a demand by progressive social movements for further democratisation, the opposite result occurred; ...the reason was that the reforms marginally eroded the position of privileged groups, thus provoking their anger, yet failed to mobilise support among less advantaged groups - women, workers, immigrants - to offset rightist opposition.

8. All these problems are well-known in the literature of state failure. As Brett (1987, p33) points out: 'The politicians and bureaucrats who control the parastatals, foreign exchange allocations, import licences and so on appropriate huge incomes in monopoly rent by manipulating access to state power'.

9. Beetham (1987, pp98-103) offers a useful discussion on why the question of democracy, or distribution of political authority is logically prior to the question of market scope. Firstly because, as we have also pointed out, the market too is a conscious historical creation. But secondly he argues that:

> the scope of the political sphere must itself be among the most important questions for political authority to decide... It is, after all, a strange concept of autonomy which accords people the right to decide what kind of shoes to buy, but allows them no part in deciding what kind of choices should in principle be available in their society, within what limits, and how they should be distributed, ie no say in what

kind of society they should live in. (Beetham, 187, p101).

10. PASOK after 1985 shifted its original anti-EC position and argued for the need for the internal market to be articulated with the 'cohesion' of the European Community. It promoted the need for the EC to expand its own funds, coordinate macroeconomic policy and carry out joint industrial and regional policies. For an account of PASOK's new pro-EC policies, see Pangalos (1987).

11. Labour's victory in 1974 followed Heath's government confrontation with the miners, which entailed that, in a context of already very strong pressure for public sector wage rises, it was difficult for Labour to resist wage demands.

Bibliography

* denotes that the reference is in Greek

Aaronovitch S (1981) <u>The Road from Thatcherism: The AES</u>, London, Lawrence and Wishart

Agapitos G (1986) * <u>The Taxation of Income in Greece</u>, KEPE, Athens, 1986

Alesina A (1988) "The End of Large Public Deficits", in Giavazzi and Spaventa eds.

Allsopp C J and Helm D (1985) "The Political Economy of Economic Policy", <u>Times Literary Supplement</u>, 6 December

Anastasakos S (1985) * "Supervisory Councils and the State", <u>Demosios Tomeas</u>, no.3, December

Anderson P (1976) <u>Considerations on Western Marxism</u>, Verso, London

Anderson P (1977) "The Anomolies of Antonio Gramsci", <u>New Left Review</u>, 100

Anderson P (1980) <u>Arguments within English Marxism</u>, Verso, London

Anderson P (1983) <u>In the Tracks of Historical Materialism</u>, London, Verso

Anderson P and Blackburn R eds. (1965) <u>Towards Socialism</u>,

Ithaca, Cornell University Press
Andrianopoulos A et al (1980) * PASOK and Power, Athens, Paratiritis
Angelopoulos A (1986) * Economic Problems, Estias, Athens
Archibugi F (1978) "Capitalist Planning in Question", in Holland (1978a) ed.
Arrow K J and Debreu G (1954) "The Existence of an Equilibrium for a Competitive Economy", Econometrica, 22, 3 (July)
Arrow K J and Hahn F H (1971) General Competitive Analysis, San Francisco, Holden-Day; Edinburgh, Oliver and Boyd
Arsenis G (1981) * in Centre for Mediterranean Studies
Arsenis G (1984) * Interview to Anti, no.289, 17/5/84
Arsenis G (1985a) * "A Stabilisation Programme without Development is a Conservative Policy", Anti, 8/11/85
Arsenis G (1985b) Interview Given to Oikonomicos Tachydromos, 1 August
Arsenis G (1986) * Speech to Parliament on Problematic Firms, in Minutes of the Parliament of the Greeks, 4th Period, 2nd Synod, Session 33, Friday 28 November, pp1544-1546
Arsenis G (1987) * Politiki Katathesi, Athens, Odysseas
Artis M J (1981) "Incomes Policies: some rationales", in Fallick and Elliot eds.
Athanasopoulos A (1985) * "Stabilisation of the Economy Through Development", Exormisi, 9/8/85
Attali J (1978) "Towards Socialist Planning", in Holland ed. (1978a)
Axt H-J (1984) "On the Way to Self-Reliance? PASOK's Government Policy in Greece", Journal of Modern Greek Studies, vol. 2, October
Babanasis S (1984) * "Methods of Education and the Goals of the Five-Year Development Plan", Oikonomikos Tachydromos, 22 November
Bank of Greece Annual Reports, 1979-88
Barrat-Brown M (1981) "International Workers' Plans" in Coates K ed.
Barrat-Brown M (1983) "Foreign Trade: Economic Planning and Industrial Democracy" in Topham T ed.
Beckerman W ed. (1972) The Labour Government's Economic Record 1964-70, London, Duckworth
Beckerman W (1974) "Labour's Plan for Industry", New Statesman, June 8

Beetham D (1985) "Surplus Value", New Socialist, no.2

Beetham D (1987) Bureaucracy, Open University Press, Milton Keynes

Bell D S ed. (1980) Labour into the Eighties, London, Croom Helm

Bergopoulos K (1986a) * "Economic Crisis and Modernisation in Greece and in the European South", in Manesis et al eds.

Bergopoulos K (1986b) * "Economic Policy in the face of the Current Recession: The Example of Southern Europe", Greek Review of Social Research, no. 63

Bernstein E (1909) Evolutionary Socialism, London, ILP Press

Best M (1982) "The Political Economy of Socially Irrational Products", Cambridge Journal of Economics, March

Best M (1986) "Strategic Planning, the New Competition and Industrial Policy", in Nolan and Paine ed.

Blackburn R (1965) "The New Capitalism", in Anderson and Blackburn eds. Blake D and Ormerod P ed. (1980) The Economics of Prosperity, London, Grant McIntyre

Blazyca G (1983) Planning is Good for You: The case for Popular Control, London, Pluto Press

Bleaney M (1983) "Conservative Economic Strategy" in Hall and Jaques eds.

Bleaney M (1985) The Rise and Fall of Keynesian Economics, London, Macmillan

Bodington S (1981) "Workers' Plan", in Coates K ed.

Boddy M and Fudge C eds (1984) Local Socialism?, London, Macmillan

Bottomore T and Goode P eds. (1978) Austro-Marxism, Clarendon Press, Oxford

Brett E A (1987) "States, Markets and Private Power in the Developing World: Problems and Possibilities", IDS Bulletin, vol. 18, no.3 (July)

Bronner S (1981) Rosa Luxemburg, Pluto Press

Brus W (1973) The Economics and Politics of Socialism, London, RKP

Burchell S and Tomlinson J (1981) The AES, Control of Capital and the Enterprise, mimeo

Burkitt B (1982) "Collective Bargaining, Inflation and Incomes Policy", in Currie D and Sawyer M eds. Socialist Economic Review 1982, London, Merlin Press

Carrillo S (1977) Eurocommunism and the State, London, Lawrence and Wishart

Castles F (1978) <u>The Social Democratic Image of Society</u>, London, Routledge
Catephores G (1983) "Greece: The Empiricist Solutions of PASOK", in M Sawyer and K Schott eds., <u>Socialist Economic Review 1983</u>, London, Merlin Press
Cecchini P (1988) <u>The European Challenge: 1992</u>, Wildwood House, UK
Centre for Mediterranean Studies (1981) * <u>The Transition to Socialism</u>, Athens, Aletri
Chamberlain E H (1933) <u>The Theory of Monopolistic Competition</u>, Harvard University Press
Clower R W (1965) "The Keynesian Counter-Revolution: a Theoretical Reappraisal" in Hahn F H and Brechling R eds., <u>The Theory of Interest Rates</u>, London, Macmillan
Coates D (1980) <u>Labour in Power?</u>, London, Longman
Coates K ed. (1980) <u>Joint Action for Jobs: A New Internationalism</u>, Nottingham, New Statesman/Spokesman
Coates K ed. (1981) <u>How to Win</u>, Nottingham, Spokesman
Cohen S (1977) <u>Modern Capitalist Planning, The French Model</u>, University of California Press
Colletti L (1972a) <u>From Rousseau to Lenin</u>, Monthly Review Press, New York and London
Colletti L (1972b) "Bernstein and the Marxism of the Second International", in Colletti (1972a)
Collis C and Turner R K (1977) <u>The Economics of Planning</u>, London, Macmillan
Commercial Bank of Greece (1982a) * "The 1982 Budget. The New Economic Policy" in <u>Economic Bulletin</u>, No.111, Jan-March
Commercial Bank of Greece (1982b) * "The Greek Economy and the New Economic Policy" in <u>Economic Bulletin</u>, No.112, April-June
Commercial Bank of Greece (1982c) * "The New Development Law 1262/82", <u>Economic Bulltin</u>, no.113, July-September
Commercial Bank of Greece (1983) * "Economic Policy During 1983" in <u>Economic Bulletin</u>, Jan-March
Commercial Bank of Greece (1985) "Public Enterprises and Entities", <u>Economic Bulletin</u>, July-September, no.125
Cooper R N (1987) "External Constraints on European Growth" in Lawrence R Z and Schultze C L eds <u>Barriers to European Growth</u>, Washington DC, The Brookings Institution
Corden W M (1981) <u>Inflation, Exchange Rates and the World</u>

Economy, 2 ed. Oxford University Press, Oxford
Corliras P (1986) "The Economics of Stagflation and Transformation in Greece" in Tzannatos ed.
Cowling K (1987) "An Industrial Strategy for Britain: The Nature and Role of Planning" International Journal of Applied Economics, vol.1
Cowling K and Sugden R (1987) Transnational Monopoly Capitalism, Sussex, Wheatsheaf Books
Crick B (1984) Socialist Values and Time, Fabian Society, no. 495
Cripps F (1981) "Government Planning as a means to Economic Recovery in the UK", Cambridge Journal of Economics
Crosland A (1956) The Future of Socialism, London, Cape
Crosland A (1974) Socialism Now and other Essays, London, Cape
Crouch ed (1979) State and Economy in Contemporary Capitalism, London, Croom Helm
Crouch (1979b) "The State, Capital and Liberal Democracy" in Crouch ed.
Crouch (1983) "Industrial Relations" in Griffith ed.
Crouch and Pizzorno A eds (1978) The Resurgance of Class Conflict in W. Europe since 1968, Volume 1, London, Macmillan
CSE London Working Group (1980) The Alternative Economic Strategy, London, CSE Books/Labour Coordinating Committee
Dearlove J and White G (1987) "Editorial Introduction to the Retreat of the State?", IDS Bulletin, vol. 18, no.3
Debreu G (1959) The Theory of Value, New York, Wiley
Dedousopoulos A (1987) * "On the Crisis" Politis, No.75(3), January
Deleau M (1987) Industrial Investment in Greece: Analysis and Recommendations, Report commissioned by OECD for the Ministry of National Economy in Athens
Delors J (1978) "The decline of French Planning", in Holland ed. (1978a)
Desai M, Auerback P and Shamsavari A (Desai et al) (1988) "The Transition from Actually Existing Capitalism", New Left Review, no. 170 (July-August)
Devine P (1974) "Inflation and Marxist Theory", Marxism Today, March
Devine P (1981) "Principles of Democratic Planning", in D Currie

and R Smith eds. Socialist Economic Review 1981, London, Merlin Press

Dolianitis D (1980) * "Intermediate Goals in the Transition to Socialism", Enimerotiko Deltio, October-December, KE.ME.DIA., PASOK

Doxiadis A (1984) * "Is there Monopoly Capitalism in Greek Industry?", in Synchrona Themata, no. 22 (July/September)

Drakos G (1986) "The Socialist Economic Policy in Greece: A Critique", in Tzannatos (ed)

Dretakis M (1987) * "I Poria tis Oikonomias kai to Mellon tou topou" Eleutherotipia, 24-27 August 1987

Durbin E (1984) "Fabian Socialism and Economic Science" in Pimlott ed.

Durbin E (1985) New Jerusalems: The Labour Party and the Economics of Social Democracy, RKP, London

Eatwell J and Green R (1984) "Economic Theory and Political Power", in Pimlott ed.

Edgley R (1982) "Revolution, Reform and Dialectic" in Parkinson G H R ed., Marx and Marxisms, Royal Institute of Philosophy Lecture Series, 14, CUP

Eichner A S (1978) "Post-Keynesian Theory: An Introduction", Challenge, May/June

Elephantis A (1981a) "PASOK and the Elections of 1977: The rise of the Populist Movement" in Penniman H R ed.

Elephantis A (1981b) * Unfound Socialism, Athens, Politis

Eleutheriou R (1983) * "Nine Years PASOK", Anti, nos. 240-43 (three parts)

Ellman M (1986) "Images of a socialist economy: the end of the statist model" in Nolan and Paine eds.

Esping-Anderson G (1985) Politics against Markets: the Social Democratic Road to Power, Princeton University Press

Estrin S (1980) "Future Planning in the UK" in Bell ed.

Estrin S and Holmes P (1983) French Planning in Theory and Practice, London, Allen and Unwin

Fallick J L and Elliot R F eds. (1981) Incomes Policies, Inflation and Relative Pay, London, Allen and Unwin

Featherstone K (1987) "PASOK and the Left" in Featherstone K and Katsoudas K eds., Political Change in Greece, London and Sydney, Croom Helm

Fine B (1980) Economic Theory and Ideology, London, Edward Arnold

Fine B and Harris L (1976) "The Debate on State Expenditure",

New Left Review, 98 (July/August)
Finn E and Levine P (1983) "Collective bargaining and Alternative Strategies", Sawyer M and Schott K eds., Socialist Economic Review 1983, London, Merlin Press
Fischer I (1933) "The Debt Deflation Theory of Great Depressions", Econometrica, October
Fleming J S (1978) "The Economic Explanation of Inflation", in Hirsch and Goldthorpe eds.
Galbraith J (1967) The New Industrial State, Boston, Houghton Mifflin
Garrett G and P Lange (1985) "The Politics of Growth", Journal of Politics, vol. 47, no. 3-4 (December)
Gerantas A (1986) * "The Supervisory Councils", Demosios Tomeas, no. 5, February
Geras N (1976) The Legacy of Rosa Luxemburg, London, Verso
Geroski P (1989) "European Industrial Policy and Industrial Policy in Europe", in Oxford Review of Economic Policy, vol. 5 no. 2
Giannitsis T (1986) * "Greece: Industrialisation in Greece", in Manesis et al (ed)
Giavazzi F and Spaventa L eds. (1988) High Public Debt: the Italian Experience, Cambridge, CUP
Giaxnis B (1985) * "Oikonomia Oratotis Miden", Anti, 299, 13 September
Gibson H D and Tsakalotos E (1990) Capital Flight and Financial Liberalisation: A study of five European countries, Univeristy of Kent Discussion Paper, no. 90/1
Giddens A and Mackenzie G eds. (1982) Social Class and the Division of Labour, Cambridge University Press
Gilhespy D et al ed. (1986) Socialist Enterprise, Nottingham, Spokesman
Glyn A (1978) Capitalist Crisis or Socialist Plan, London, Militant
Glyn A (1985) A Million Jobs a Year: The case for Planning Full Employment, London, Verso
Glyn A and Harrison J (1980) The British Economic Disaster, London, Pluto Press
Goldthorpe J H (1978) "The current inflation: towards a sociological account" in Hirsch and Goldthorpe eds.
Goldthorpe J H ed. (1984) Order and Conflict in Contemporary Capitalism, Oxford, Clarendon Press
Goldthorpe J H (1984b) "The End of Convergence: Corporatist

and Dualist Tendencies in Modern Western Societies", in Goldthorpe ed.

Goldthorpe J H (1985) Problems of Political Economy after the Post-war Period, Mimeo, subsequently published in C S Mares ed. <u>Changing Boundaries of the Political: essays on the evolving balance between the state and society, public and private in Europe</u>, 1987, Cambridge University Press

Grahl J (1981) "Discussion: Government, Trade Unions and Inflation" in D Currie and R Smith eds., <u>Socialist Economic Review 1981</u>, London, Merlin Press

Grahl J and Rowthorn B (1986) "Dodging the Taxing Questions", <u>Marxism Today</u>, November

Greek Government Programme (1981), General Secretarial for Press and Information, Athens

Griffith J (1983) <u>Socialism in a Cold Climate</u>, London, Unwin Paperbacks

Griffith-Jones S (1981) <u>The Role of Finance in the Transition to Socialism</u>, London, Frances Pinter

Grossman H and Stiglitz J E (1976) "Information and Competitive Price Systems", <u>American Economic Review</u>

Hadjigregoriou N (1979) * "Underdevelopment and Industrialisation", <u>Miniaio Enimerotiko Deltio</u>, September, KE.ME.DIA., PASOK

Hadjisocrates D (1982) * "The New Law for the Provision of Incentives 1262/82", <u>Synchrona Themata</u>, no.15, September

Hahn F (1984) Equilibrium and Macroeconomics, Basil Blackwell Hahn F (1985) "Recognizing the Limits", <u>Times Literary Supplement</u>, December 6, p4

Halikias D (1987) <u>Financial Reform and Problems of Monetary Policy</u>, Speech given to the Greek-American Chamber of Commerce, 24 November

Halikias D (1988) <u>The Problem of the Greek Balance of Payments</u>, Speech given to the Greek-German Chamber of Commerce, 25 February

Hall P (1986) <u>Governing the Economy</u>, Polity Press, Cambridge

Hall S (1983) "The Great Moving Right Show" in Hall and Jacques eds.

Hall S and Jacques M (1983) <u>The Politics of Thatcherism</u>, Lawrence and Wishart, London

Harding N (1983) <u>Lenin's Political Thought</u>, London, Macmillan

Hare P (1983) "The Preconditions for Effective Planning in the UK", in D Currie and M Sawyer eds., <u>Socialist Economic</u>

Review 1983, London, Merlin Press
Hare P (1985) Planning the British Economy, London, Macmillan
Harrison J (1982) "A 'Left' Critique of the AES", in D Currie and M Sawyer eds., Socialist Economic Review 1982, London, Merlin Press
Hatfield M (1978) The House the Left Built, London, Gollancz
Henley A and Tsakalotos E (1992) "Corporatism, Conflict and Income Distribution", Cambridge Journal of Economics, forthcoming
Hicks J R (1937) "Mr Keynes and the 'Classics": A Suggested Interpretation", Econometrica, vol. V, pp147-189
Hicks J R (1974) The Crisis of Keynesian Economics, Oxford, Basil Blackwell
Higgins W and Nixon A (1983) "How Limited is Reformism? A Critique of Przeworski and Panitch", Theory and Society, no.12, pp603-30
Himmelstrand U (1981) Beyond Welfate Capitalism, London, Heineman
Hirsch F (1977) Social Limits to Growth, London, RKP
Hirst P (1982) "The division of labour, incomes policy and industrial democracy", in Giddens and MacKenzie eds.
Hodgson G (1977a) Socialism and Parliamentary Democracy, Spokesman, Nottingham
Hodgson G (1977b) "A Reply to Shirley Williams", in Hodgson (1977a)
Hodgson G (1977c) "The Antinomies of Perry Anderson", in Hodgson (1977a)
Hodgson G (1981) Labour At the Crossroads, Oxford, Martin Robertson
Hodgson G (1982) "On the Political Economy of Socialist Transition", New Left Review, 133
Hodgson G (1984) The Democratic Economy, Penguin Books
Hodgson G (1986) "The Limits to Keynes", in Nolan and Paine eds.
Hodgson G (1988) Economics and Institutions, Polity Press, Cambridge
Holland S (1972) The State as Entrepreneur, London, Weidenfeld and Nicolson
Holland S (1975) The Socialist Challenge, London, Quartet Books
Holland S ed. (1978a) Beyond Capitalist Planning, Oxford, Basil Blackwell

Holland S (1978b) "Planning Disagreements" in Holland, ed. (1978a)
Holland S (1979) "The New Communist Economics", in Filo Della Torre et al ed., Eurocommunism: Myth or Reality
Holland S (1981) "New Public Enterprise and Economic Planning" in K Coates, ed.
Holland S ed. (1983) Out of Crisis: A Project for European Recovery, Spokesman, Nottingham
Hughes A (1986) "Investment Finance, Industrial Strategy and Economic Recovery", in Nolan and Paine eds.
Hughes J (1983) "Democratic Planning for an Enfeebled Economy", in Topham T ed.
Hunt E K and Schwartz J G (1972) A Critique of Economic Theory
Jessop B (1982) The Capitalist State, Oxford, Basil Blackwell
Kalecki M (1972) "Political Aspects of Full-employment" in Hunt and Schwartz eds.
Kapsis P (1983) * "The Economic Policy of 1982" Sinchrona Themata, No.16, February
Katseli L (1984) * "Five-Year Plan and Investment Policy", paper given to Conference organised by the Agricultural Bank of Greece, 27 June
Katseli L (1985a) * "The Economy", Enimerotiko Deltio, January-February, KE.ME.Dia, PASOK
Katseli L (1985b) "Greek Experience under PASOK: Lessons for Development Policy", lecture delivered at the Royal Institute of International Affairs, 13 November 1985
Katseli L (1985c) * "Contradictions of short-term and Development Policy", lecture delivered at IMEO, 27 November 1985.
Katseli L (1986) "Building a Process of Democratic Planning", in Tzannatos ed.
Katseli L (1990) Structural Adjustment of the Greek Economy, CEPR Discussion Paper, no. 374, February
Keat R and Urry J (1982) Social Theory as Science, London, RKP
KEPE (1987) The Five-Year Economic and Social Development Plan 1983-87: A Summary, Athens, Centre of Planning and Economic Research
Kesselman M (1986) "Lyrical Illusions or a Socialism of Governance: Whither French Socialism?", in Miliband et al eds., Socialist Register 1985/86, Merlin Press
Keynes J M (1931) Essays in Persuasion, The Collected Writings

of J M Keynes, vol.IX, London, Macmillan
Keynes J M (1964) The General Theory of Employment, Interest and Money, New York, Harvest Books
Kitching G (1983) Rethinking Socialism, Methuen
Kolmer K (1986) * Distixos Eptoxeusamen, Roes, Athens
Kopanitsa D (1986) * "How to Restructure Problematic Firms" Oikonomicos Tachydromos, 21 August
Korliras P (1986) "The Economics of Stagflation and Transformation in Greece" in Tzannatos (ed)
Korpi W (1978) The Working Class in Welfare Capitalism, Routledge, London
Korsch K (1975) "What is Socialization?", New German Critique, vol. 6, pp60-82
Kotzias N (1984) * The 'Third Road' of PASOK, Athens, Odysseas
Koulouglou S (1986) * In the Tracks of the Third Road, Odysseas, Athens
Laclau E and Mouffe C (1981) "Socialist Strategy - Where Next?", Marxism Today
Lambrianidis L (1987) * "The Planning of Development", Politis, vol. 83, no. 12 (October)
Landesmann M (1986) "UK Policy and the International Economy: an Internationalist Perspective" in Nolan and Paine eds.
Lane D (1981) Leninism: A sociological interpretation, Cambridge, CUP
Lane T (1987) "Unions: Fit For Active Service", Marxism Today, February
Lange O (1935) Marxian Economics and Modern Economic Theory", Review of Economic Studies, vol. III, no.3
Lange O (1938) On the Economic Theory of Socialism, University of Minnesota Press
Lazaris A (1981) * in Centre for Mediterranean Studies
Lehmbruch G (1984) "Concertation and the Structure of Corporatist Networks" in Goldthorpe ed.
Leijonhufvud A (1968) On Keynesian Economics and the Economics of Keynes, Oxford, OUP
Lekatsa E (1986) * "The Liberalisation of the Banking System is a Precondition for the Reconstruction of Problematic Firms", Oikonomikos Tachydromos, 21 August
Leser N (1976) "Austro-Marxism: A reappraisal", Journal of Contemporary History, vol. 11, pp133-48
Levacic R and Rebman A (1984) Macroeconomics, London,

Macmillan

Levine A (1984) <u>Arguing for Socialism</u>, RKP, London

Leys C (1985) "Thatcherism and British Manufacturing" <u>New Left Review</u>, No. 151 (May-June)

Lichtheim G (1974) <u>A Short History of Socialism</u>, London, Verso

Lindblom C E (1977) <u>Politics and Markets: The World's Political-Economic Systems</u>, New York, Basic Books

Lipsey D (1982) "A 'Right' Critique of the AES" in Currie D and Sawyer M, <u>Socialist Economic Review 1982</u>, London Merlin Press

Lukacs G (1971) <u>History and Class Consciousness</u>, London, Merlin Press

Lukes S (1984) "The Future of British Socialism", in Pimlott ed.

Lutz V (1969) <u>Central Planning for the Market Economy</u>, London, Longman

Luxemburg R (1970) <u>Reform or Revolution</u>, New York, Pathfinder Press

Lyberaki A (1988) <u>Small Firms and Flexible Specialisation in Greek Industry</u>, PhD Thesis, University of Sussex

Marglin S and Schor J eds (1990) <u>The Golden Age of Capitalism</u>, Clarendon, Oxford

McFarlane B (1984) "Economic Planning, Past Trends and Future Prospects", <u>Contributions to Political Economy</u>, vol.3 (March)

McLennan G, Held D and Hall S eds. (1984) <u>State and Society in Contemporary Britain</u>, Cambridge, Polity

Malinvaud E (1977) <u>The Theory of Unemployment Reconsidered</u>, Basil Blackwell, Oxford

Manesis A et al (1986) * <u>Greece in Development</u>, Exantas, Athens

Marinos-Kouris D (1985) "The Industrial Policy by Branch of Industry", <u>Economic Bulletin</u>, no. 125, July-September, Commercial Bank of Greece

Martin A (1979) "The dynamics of change in a Keynesian political economy: the Swedish case and its implications", in Crouch ed.

Martin D (1988) <u>Bringing Common Sense to the Common Market: A Left Agenda for Europe</u>, Fabian Society, no. 525

Masse P (1962) "French Methods of Planning", <u>Journal of Industrial Economics</u>, vol.XI, no.1

Masse P (1965) "Economic Planning and Economic Theory",

Econometrica, vol.33
Matthews R C O (1968) "Why has Britain had Full-Employment Since the War" Economic Journal, vol. 78 no.3
Meacher M (1981) "The Alternative: Can it Work?" in K Coates ed.
Meade J (1970) The Theory of Indicative Planning, Manchester University Press
Medvedev R (1981) Leninism and Western Socialism, London, Verso
Meidner R (1978) Employment Investment Funds, London, Allen and Unwin
Minns R (1982) Take Over the City, London, Pluto Press
Mitrafanis G N (1985) * "When Informal Procedures Replace Institutions", Politis, no. 45, 12 July
Mitropoulos A (1985) * Labour Relations in Greece, Athens, Sakkoula
Mitropoulos A (1986) * "Socialisation in Greece", Sindicalistiki Epitheorisi, no. 22, October
MNE (1985) Stabilisation of the Economy, Athens, Ministry of National Economy
MNE (1986) * Report on the Progress of the Five-Year Plan for Economic and Social Development for the Period 1983-85, Athens, Ministry of National Economy
MNE (1987) * DEKO: Conference on Socialisation, Athens, Ministry of National Economy
MNE/KEPE (1985) * Plan for Economic and Social Development 1983-87, MNE/KEPE, Athens, August
Modigliani F (1944) "Liquidity Preference and the Theory of Money", Econometrica, vol.XII, pp48-88
Modinos M (1983) * "Five-Year Plan and Popular Participation but Planning has been Forgotten", Dekapenthimeros Politis, no. 5, 30 December
Mouzelis N (1978) Modern Greece: Facets of Underdevelopment, Macmillan, London
Mouzelis N (1980) * "Thoughts on the Impressive Political Rise of PASOK", in Andrianopoulos et al eds.
Murray R (1984) "New Directions in Municipal Socialism", in Pimlott ed.
Murray R (1985) "The New Economic Order", Marxism Today
Murray R (1987) "Ownership, Control and the Market", New Left Review, No. 164, July-August
Negroponti-Delivani M (1986) * The Problematic Greek Industry

and Some Solutions, Athens, Paratiritis

Nikolinakos M (1984) * Problematic Firms, Socialisations, Supervisory Councils and the Role of the Private Sector, Conference paper, Athens, IMEO

Nolan P and Paine S eds (1986) Rethinking Socialist Economics, Polity Press

Nove A (1983) The Economics of Feasible Socialism, Allen and Unwin, London

Nuti M (1986) "Economic Planning in Market Economies: Scope, Instruments, Institutions", in Nolan and Paine eds.

Odling-Smee J and Riley C (1985) "Approaches to the PSBR" National Institute Economic Review, no. 113

OECD Economic Surveys: Greece, 1979 to 1987

Offe C and Ronge V (1984) "Theses on the Theory of the State" in Giddens A and Held D eds., Classes, Power and Conflict, London, Macmillan

Ormerod P (1981) "Inflation and Incomes Policy" in D Currie and R Smith eds., Socialist Economic Review, London, Merlin

Ormerod P and Blake D (1980) The Economics of Prosperity: Social Priorities in the Eighties, London, McIntyre

Palma G (1978) "Dependency: A formal theory of underdevelopment or a methodology for the analysis of concrete situations of underdevelopment", World Development, vol.6, pp881-924

Panagopoulos S (1985) * "Towards a New Policy for Reconstruction of Problematic Firms", Biomichaniki Epitheorisi, 33(389), November - December

Pangalos T (1987) * "European Integration and the Left" Socialist Theory and Practice, 4/87 (August), PASOK

Panitch L (1981) "Trade Unions and the capitalist state", New Left Review, no. 125

Panitch L (1986) "The Impasse of Social Democratic Politics", Socialist Register, 1985/86

Papadatos G (1984) * Issues in the Transition to Socialism, Athens, Aihmi

Papagiannakis L (1988) * "Industrialisation: The Continually Unfulfilled Dream of Modern Greece", Oikonomikos Tachydromos, 3 March

Papagiannakis M (1980) * "PASOK from Opposition to Government" in Andrianopoulos et al eds.

Papamichail A (1978) * "Some Problems in the Stage of the Transition to Socialism", Enimerotiko Deltio, April-May,

KE.ME.DIA., PASOK

Papamichail A (1982) * "Problems of Power in the Transition to Socialism", Enimerotiko Deltio, 2/82, September, KE.ME.DIA., PASOK

Papamichail A (1983) * "The Strategy of Democratic Rupture", Enimerotiko Deltio, 3/83, August, KE.ME.DIA., PASOK

Papandreou A (1971) Paternalistic Capitalism, University of Minnesota Press, Minneapolis; London, Oxford University Press

Papandreou A (1976) * Greece for the Greeks, Athens, Karanasi

Papandreou A (1977) * For a Socialist Society, Athens, Aihmi

Papandreou A (1978) * The Transition to Socialism, Athens, Aihmi

Papandreou A (1978b) * "Speech on the Budget (1978)", Miniaio Enimerotiko Deltio, May, KE.ME.DIA, PASOK

Papandreou A (1979) * "Speech on the Budget (1979)", Enimerotiko Deltio, January, KE.ME.DIA., PASOK

Papandreou A (1980) * "Speech on the Budget (1980)", Poria Gia Tin Allaghi, KE.ME.DIA.,A 26/1981, PASOK

Papandreou A (1981a) * in Centre for Mediterranean Studies (1981)

Papandreou A (1981b) * Interview in Exormisi, 30 August

Papandreou A (1982) * Speech to Thessaloniki International Trade Fair, Macedonia, 12 September

Papandreou A (1983a) * Speech to Thessaloniki International Trade Fair, Macedonia, 11/9/83

Papandreou A (1983b) * Interview given to Eleutherotipia, 17/11/83

Papandreou A (1983c) * Speech to 10th Synod of PASOK's Central Committee, in PASOK (1983b)

Papandreou A (1984) Speech to PASOK Congress, 10 May, Athens, KE.ME.Dia, PASOK

Papandreou A (1985a) * Speech to the 15th Synod of PASOK's Central Committee, Oikonomikos Tachydromos, 14/2/85

Papandreou A (1985b) * Speech to Thessaloniki International Trade Fair, To Vima, 1/9/85

Papandreou A (1986) Speech to 1st Special Synod of PASOK's Central Committee on Economic Policy and Development, Theoria kai Praxis, 1/86

Papandreou V (1986a) * Speech to Parliament on Problematic Firms, Minutes of the Parliament of the Greeks, 4th Period, 2nd Synod, Session 33, Friday 28 November, pp1547-1550

Papandreou V (1986b) * "The Development Plan for Industry and Handicrafts", Theoria kai Praxis, no.1/86

Papandropoulos A X (1985) * "What is the role of the National Council of Development" Oikonomicos Tachydromos, 15 July

Papantoniou G (1981) * in Centre for Mediterranean Studies

Papantoniou G (1986) * Paper delivered at Conference organised by Internatonal Herald Tribune, Athens 12/5/86

Papantoniou G (1987a) * Speech to Parliament in January 1987 on the Five-Year Plan (1983-87), Athens, MNE

Papantoniou G (1987b) * "Planning and Development", Oikonomicos Tachydromos, 8 October

Parkin F (1979) Marxism and Class Theory: A Bourgeois Critique, Tavistock, London

PASOK (1974) * "Declaration of the Third of September", reprinted in Eleutherotipia, 4 and 5 September

PASOK (1975) * Socialist Transformation, KE.ME.DIA., PASOK

PASOK (1977) * General Guidelines for Government Policy, Athens, PASOK

PASOK (1981) * Contract with the People, Declaration of Government Policy, KE.ME.DIA., PASOK

PASOK (1982a) * National Popular Unity: second synod of PASOK central committee, Third Edition, KE.ME.DIA., PASOK

PASOK (1982b) People-PASOK, The Road of Allaghi: ninth session of PASOK central committee, First Edition, KE.ME.DIA., PASOK

PASOK (1983a) * "The Role and Importance of Institutions", Enimerotiko Deltio, 2/83, July, KE.ME.DIA., PASOK

PASOK (1983b) * Institutional Changes and Socialist Transformation: Tenth Synod of PASOK Central Committee, KE.ME.DIA, PASOK

PASOK (1986) * "Issues on the nature and role of social democracy", Enimerotiko Deltio, July-August, KE.ME.DIA., PASOK

PASOK Working Team (1984) * "Economic Policy: The Budget" Enimerotiko Deltio, Jan. 1984, KE.ME.DIA, PASOK

Patinkin D (1956) Money, Interest and Prices, New York, Harper and Row

Paulopoulos P (1986) * Income Shares: Tendencies, Causes, Effects, IOBE, Athens

Penniman H R (1981) Greece at the Polls - The National Elections of 1974 and 1977, Washinton D C, American Enterprise Institute

Perrakis C (1985) "Development Aspects of the Five-Year Plan", Economic Bulletin, no. 125, July-September, Commercial Bank of Greece

Pesaran M Hashem (1986) "Structural Keynesianism as an alternative to Monetarism" in Nolan and Paine (eds)

Petmetzidou-Tsoulouvi M (1984) * "Approaches to the Issue of Underdevelopment in the Greek Social Formation: A Critical Review" Synchrona Themata, No. 22, (July/September)

Petmetzidou-Tsoulouvi M (1987) * Social Classes and Mechanisms of Social Reproduction, Exantas, Athens

Petras J (1983) "Problems in the Transition to Socialism", Monthly Review, May

Pimlott B ed. (1984) Fabian Essays in Socialist Thought, London, Heineman

Pizzorno A (1978) "Political Exchange and Collective Identity in Industrial Conflict", in Crouch and Pizzorno eds.

Plant R (1984) Equality, Markets and the State, Fabian Society, no. 494

Pontusson J (1984) "Behind and Beyond Social Democracy in Sweden", New Left Review, 143 (January/February)

Pontusson J (1987) "Radicalisation and Retreat in Swedish Social Democracy", New Left Review, no.165 (September/October)

Poulantzas N (1973) Political Power and Social Classes, London, New Left Books

Poulantzas N (1980a) State, Power, Socialism, London, Verso

Poulantzas N (1980b) * "Is the Unity of the Forces of Allaghi Possible?" in Andrianopoulos et al

Prior M ed (1981) The Popular and the Political

Prior M and Purdy D (1979) Out of the Ghetto

Przeworski A (1985) Capitalism and Social Democracy, Cambridge, CUP

Purdy D (1977) "Review Article", Capital and Class, no.1

Purdy D (1981b) "The Social Contract and Socialist Policy" in Prior ed.

Radice H (1984) "The National Economy: A Keynesian Myth?" Capital and Class, Spring, pp111-140

Rawls J (1971) A Theory of Justice, Cambridge, Mass., Harvard University Press

Regini M (1984) "The conditions for Political Exchange: how Concertation emerged and collapsed in Italy and Great Britain" in Goldthorpe ed.

Richardson G (1971) "Planning vs. Competition", <u>Soviet Studies</u>, vol.2

Robinson J (1933) <u>The Economics of Imperfect Competition</u>, London, Macmillan

Robinson J (1975) "What has become of the Keynesian Revolution", in M Keynes ed. <u>Essays in Honour of J M Keynes</u>

Roemer J (1982) <u>A General Theory of Exploitation and Class</u>, Cambridge Mass., Harvard University Press

Rowthorn B (1980) <u>Capitalism, Conflict and Inflation</u>, Lawrence and Wishart, London

Rowthorn B (1981) "The Politics of the AES", <u>Marxism Today</u>, (January)

Rowthorn B (1983) "The Past Strikes Back", in Hall and Jaques eds.

Salvadori M (1979) <u>Karl Kautsky and the Socialist Revolution</u>, New Left Books, London

Sambethai I D (1986) <u>Report on the Greek Labour Market</u>, Economic Reseach Department, Bank of Greece

Sawyer M (1985) <u>The Economics of Michal Kalecki</u>, London, Macmillan

Schott K (1982) "The rise of Keynesian Economics: Britain 1940-69", <u>Economy and Society</u>, vol.11, no.2

Schumpeter J (1942) <u>Capitalism, Socialism and Democracy</u>, New York, Harper and Row

Scitovsky T (1978) "Market Power and Inflation", <u>Economica</u>, 45

Sharples A (1981) "Alternative Economic Strategies", in D Currie and R Smith eds., <u>Socialist Economic Review 1981</u>, London, Merlin Press

Sheriff T (1980) "Trade, Industry and Employment" in Blake and Ormerod eds.

Shonfield A (1969) <u>Modern Capitalism</u>, Oxford, OUP

Simitis K (1986) * "Directions for Development Policy", <u>Theoria kai Praxis</u>, no.1/86

Simitis K (1987) * Speech to Parliament on Planning Agreements and Development Contracts on 19 January, Athens, MNE

Skidelsky R (1977) "The Political Meaning of the Keynesian Revolution", in Skidelsky R ed. <u>The End of the Keynesian Era</u>, London, Macmillan

Skidelsky R (1979) "The Decline of Keynesian Politics" in Crouch ed.

Spaventa L (1988) "Introduction: is there a public debt problem

in Italy", in Giavazzi and Spaventa eds.
Spourdalakis M (1986) "The Greek Experience" in Miliband et al, eds. Socialist Register 1985-86, Merlin Press
Spraos J (1984) * "Mia Helliniki Idiomorphia", Ta Nea, 9/8/84
Spraos J (1986) Government and the Economy: the First Term of PASOK 1982-84, Unpublished Paper
Stalin J (1953) The Foundations of Leninism, Vol. 6, Foreign Languages Editions
Steindl J (1985) "J M Keynes, Society and the Economist", in Vicarelli F ed.
Stephens J (1979) The Transition from Capitalism to Socialism, London, Macmillan
Stergiou D (1985) * "Report on Problematic Firms", Oikonomicos Tachydromos, 9 May, 4 June, 11 June, 18 June, 25 June
Stiglitz J E (1984) "Information and Economic Analysis: A Perspective", Economic Journal, Conference Papers, pp21-40
Stiglitz J E (1986) The Economics of the Public Sector, New York, Norton
Stournaras Y (1987) * The Financial Restructuring of Ailing DEKO (public enterprises and corporations), Ministry of the National Economy
Sweezy P M (1940) The Theory of Capitalist Development, New York, Monthly Review Press
Thirlwall A P (1986) Balance of Payments Theory and the United Kingdom Experience, London, Macmillan
Thompson E (1960a) "Revolution", New Left Review, 3
Thompson E (1960b) "Revolution Again", New Left Review, 6
Thompson G (1984) "Economic Intervention in the Post-war Economy" in McLennan et al, eds.
Tobin J (1972) "Inflation and Unemployment", American Economic Review, vol 62
Tobin J (1980) Asset Accumulation and Economic Activity, Oxford, Basil Balckwell
Tomlinson J (1982) The Unequal Struggle? British Socialism and the Capitalist Enterprise, London, Methuen
Tomlinson J (1988) Can Governments Manage the Economy?, Fabian Society, no. 524
Topham T ed (1983) Planning the Planners: How to Control the Recovery, Nottingham, Spokesman
Tsotsoros S N (1985) * Problematic Firms and the Intervention of Law 1386/83, Athens, OAE
Tsoukalas C (1986) "Radical Reformism in a 'Pre-welfare"

Society: the Antinomies of Democratic Socialism in Greece" in Tzannatos (1986) ed.

TUC/Labour Party Liason Committee (1982) *Economic Planning and Industrial Democracy*

Tzannatos Z ed (1986) *Socialism in Greece*, Gower, London

Vaitsos C (1986) "Problems and Policies of Industrialisation" in Tzannatos ed.

Valden S (1985) * "The Economic Measures", *Aristera Simera*, no. 14

Vicarelli F ed. (1985) *Keynes' Relevance Today*, London, Macmillan

Wainwright H and Elliot D (1982) *The Lucas Plan: A New Trade Unionism in the Making*, London, Allison and Busby

Ward T (1981) "The case for Import Control Strategy in the UK, in D Currie and R Smith eds., *Socialist Economic Review 1981*, London, Merlin Press

Wilson A and Green R (1982) "Economic Planning and Workers' Control" in M Eve and D Musson eds., *Socialist Register 1982*, London, Merlin Press

Wilson A and Green R (1983) *The Future Course of Planning*, in Topham ed.

Wright A (1984) "Tawneyism Revisited. Equality, Welfare and Socialism", in Pimlott ed.

Xanthakis M D (1985) "The Development Law 1262/82 and the Incentives to Manufacturing", *Economic Bulletin*, No. 125, July-September, Commercial Bank of Greece

Xtouris S (1987) * "Economic Crisis and Reform of the State: the limits of economic policy", *Politis*, vol. 86, no. 4 (December)

Zachariadis D (1980a) * "Dependency: Theoretical Base and Practical Choices", *Enimerotiko Deltio*, 10-12, October-November, KE.ME.DIA., PASOK

Zachariadis D (1980b) * "The Capitalist Development of Greek Industry", *Exormisi*, 22 November